Rings

Rings

Discoveries from Galileo to Voyager

*James Elliot
and
Richard Kerr*

The MIT Press
Cambridge, Massachusetts
London, England

This book was set in Baskerville by The MIT Press Computergraphics Department and printed and bound by Halliday Lithograph in the United States of America.

Library of Congress Cataloging in Publication Data

Elliot, James, 1943–
    Rings: discoveries from Galileo to Voyager.

    Bibliography: p.
    Includes index
    1. Planetary rings. I. Kerr, Richard. II. Title.
QB603.R55E44 1984   523.4   84-9721
ISBN 0-262-05031-5

Contents

Preface

This is a story of scientific discovery. You will not find it in a textbook. Textbooks will tell you that from 1977 to 1981 the field of planetary rings experienced an explosion of new discoveries and new understanding with few parallels in the history of exploration. That is true. Until 1977, Saturn's three broad rings made it *the* ringed planet. Within 4 years, astronomers discovered two unsuspected ring systems and transformed the apparently immutable three rings of Saturn into thousands of dazzling, often mysterious, ringlets. The discoveries likewise transformed the traditional questions of ring studies—Why does only Saturn have rings? Why is there a gap in the middle of them?—into a seemingly endless string of perplexing new problems. How can rings be 1,000 times narrower than any known in the preceding 300 years? What could put a kink in a ring or make it elliptical? How can a ring be only a millionth the age of the rest of the solar system? These mind-stretching discoveries precipitated a revolution, one that liberated rings from the suffocating stereotype that had circumscribed their study for centuries.

The story that will probably not appear in textbooks is the seemingly "unscientific" way that science sometimes makes progress. In the case of planetary rings, theory did not lead the way to observation and discovery—rather discovery led theory; the uniqueness of Saturn's rings had become the common wisdom and remained so for centuries. Because astronomers had such low expectations, discoveries fell to unsuspecting observers intent on business of their own. Clues to the presence of rings went unnoticed or were put aside as useless. Isolated bits of information that, if taken together, might have directed observers to new discoveries, remained unsynthesized and impotent when the lines of scientific communication broke or never formed at all. Discoverers failed, at first, to recognize their own discoveries for the strangeness of their garb.

If the exploration of planetary rings is any example, the process of discovery is far more complex and often more circuitous than the name-date-place style of textbooks would imply. We know it is more interesting than that. We present this account of ring discoveries in the hope that the reader, whether layman or scientist, may better appreciate the nature of scientific progress. It is not presented in the form of a scientific textbook or a scholarly history; rather the story includes sufficient technical detail to trace the paths to discovery as well as to convey some of the wonder of planetary rings. Thus, there is no rigorous analysis of theory or observation. These can be found in the references in the bibliography. As in any story of science, perceptions of the truth change with time. We have not pointed out every false turn as it happened, only tried to carry the reader to our understanding of rings at the time we wrote the manuscript.

In chapter 1, we present a portion of the transcript of conversations aboard the Kuiper Airborne Observatory during the discovery of the Uranian rings. Although not as effective as listening to the tape (copies are available from The MIT Press), it gives the flavor of the changing thoughts and attitudes of the participants as the incoming data forced old ideas to be abandoned in favor of the new ones. Preparing the transcript proved not so straightforward as listening to the tape and writing down exactly what was said. The format of written communication differs from that of verbal communication—people do not always speak in complete sentences, one person at a time, and certain words become packed with special meanings for those who have been working closely together on a project. In addition, the high noise level and automatic gain control of the intercom system on the airplane added their own difficulties. Just a few weeks after the event, Mary Roth transcribed the tape. Then Ted Dunham and J.E. listened to the tape and went over the transcript, filling in the names of speakers and clarifying unintelligible conversation in preparation for an article written at the time for *Sky and Telescope*. Even with two participants going over the tape several times, only a few weeks after the event, uncertainties remain in exactly what was said, not to mention exactly what was meant.

We bring two special elements to the story—J.E.'s firsthand experience with rings (the first person of chapters 1, 3, 4, 10, and 11 reflects his role in the discovery of the Uranian rings and his further work in the field) and R.K.'s coverage of Jovian and Saturnian ring discoveries for *Science* magazine. In addition, a large number of scientists and others generously assisted us in gathering further information. We thank them and in particular mention Joe Churms, Jeffrey Cuzzi,

Ted Dunham, Larry Esposito, Walter Feibelman, Carl Gillespie, Peter Goldreich, Bob Millis, David Morrison, Phil Nicholson, Tobias Owen, and Scott Tremaine. We also thank Lyn Elliot, Martha Elliot, Marianne Connolly, and Heidi Hammel for helping to prepare the manuscript.

Richard Kerr
James Elliot

Rings

1

Odd Flickers from a Star

Heavy with a full 153,000-pound load of fuel, the C-141 left the runway of the Perth International Airport at 10:37 P.M. on 10 March 1977 and headed southwest. Not an ordinary course from Perth, but this was not an ordinary C-141. It was NASA's Gerard P. Kuiper Airborne Observatory, KAO for short, named after the father of modern planetary astronomy. Peering out a hole in the port side of the KAO was a 0.9-meter telescope, modest in size by today's standards, that we would use once we reached the 41,000-foot cruising altitude of the airplane. There we would be above most of the water vapor that blocks infrared radiation of astronomical objects from reaching ground-based telescopes, even those on the highest mountains. Since many objects, such as stars in their formative stages, emit most of their energy in the infrared, the KAO provides a convenient window on the universe that can otherwise be obtained only with artificial Earth satellites or high-altitude balloons.

The KAO was not out for an infrared view of star birth that night; rather we were heading toward a rendezvous with a star shadow cast by the planet Uranus. In just 6 hours, Uranus was scheduled to pass in front of a star known only by its catalog designation, SAO 158687. Such an occultation is analogous to a solar eclipse, with the star playing the role of the Sun and Uranus playing that of the Moon. This was the first occultation ever predicted for Uranus, and from it we would have a unique opportunity to learn the temperature of the Uranian upper atmosphere and to measure the precise diameter of the planet. Without the occultation opportunity, we would have had to wait until a Voyager spacecraft reached Uranus in 1986—if NASA decided to attempt the trip and the spacecraft remained healthy.

Since observations were not scheduled to begin for several hours, all aboard had plenty of time to prepare. On the flight deck, Ron

Figure 1.1
The Kuiper Airborne Observatory. Within the body of a C-141 aircraft, a type normally used for cargo, NASA has mounted a telescope that has a primary mirror 0.9 meter in diameter. At a normal cruising altitude of 41,000 feet, the primary mission of the Kuiper Airborne Observatory is to make astronomical observations in the far infrared—radiation that is strongly absorbed by water vapor in the Earth's lower atmosphere. The observatory flies about 80 astronomical missions per year. [Courtesy of NASA]

Gerdes occupied the left seat as the pilot in charge of the airplane, with copilots Dave Barth and Bob Innis available to take their turns at the controls during the scheduled 10 1/2-hour flight. Flight engineer Frank Cosik was keeping track of fuel consumption and proper functioning of aircraft systems, which would be even more critical than usual as we pushed the KAO to the safe limit of its endurance over this deserted part of the ocean. Navigator Jack Kroupa had laid out a course that would give us the maximum time in Uranus's shadow as its northern edge raced somewhere between Australia and Antarctica. Kroupa would normally have left the on-board navigation to the two inertial navigation systems, but came along because of the critical nature of the mission.

Keeping an airborne telescope aimed precisely at a star requires some sophisticated hardware; precise pointing of the telescope is

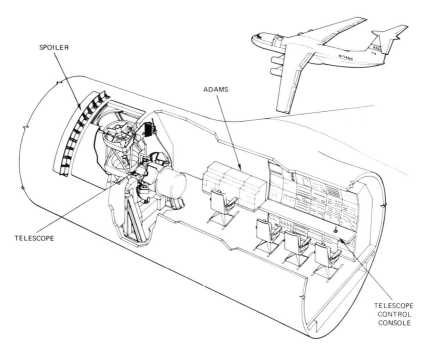

SPOILER

ADAMS

TELESCOPE

TELESCOPE
CONTROL
CONSOLE

Figure 1.2
A telescope in the sky. Mounted in a cavity open to the outside, the telescope in
the Kuiper Airborne Observatory rests on a cushion of air flowing through a
spherical bearing that isolates it from the roll, pitch, and yaw of the airplane.
Gyros keep the telescope drift rate low, and a feedback signal from the small,
star-tracking telescope keeps the main telescope pointing in a constant direction.
The spoiler influences the airflow pattern to prevent violent pressure oscillations
from building up within the open cavity. The onboard computer, ADAMS (Airborne Data and Management System), records experimental data and can be
used to control the operation of the telescope. A quartz window separates the
telescope cavity from the cabin, so that the working environment within the
plane is kept at a pressure equivalent to atmospheric pressure at 8,500 feet—an
altitude lower than many ground-based observatories. [Courtesy of NASA]

achieved through a three-step process. After the pilot trims the plane
to fly the intended course, the autopilot provides stability to about 1
degree. Next, gyros mounted on the telescope provide a feedback
signal to motors that compensate for airplane motion by moving the
telescope on its spherical air bearing. Finally, a star tracker, a small
telescope mounted on the main telescope, provides an additional feedback signal to the motors—the final pointing is usually reliable to 1
arc-second, the angle subtended by your fist at a distance of 10 miles.
That night Don Olson and Milo Reisner were responsible for operating

the telescope, with Don Oishi along to help cover unexpected problems; Al Meyer would set up and run the star tracker.

Aft of the telescope systems, Tom Matheson operated the computer, and Pete Kuhn had his infrared radiometer that monitors the amount of water vapor in the atmosphere above the plane. His measurements could warn us of clouds. Also in the rear of the plane were three guests: two flight controllers from Perth and NASA's representative in Australia, Wilson Hunter. It was NASA's job to get the plane to the prearranged observing spot for the occultation and to point the telescope at Uranus so that the light would reach our instrument mounted on the telescope. Mission directors Carl Gillespie and Jim McClenahan were coordinating these efforts.

Then it was our job as the astronomy team to get the data. As the team leader, I had proposed to NASA over a year before to observe this occultation. Along with me were my new programmer, Doug Mink, who had recently finished his master's degree at MIT, and my graduate student, Ted Dunham, who was planning to base a major portion of his doctoral research on this event. Ted would be operating the data-recording equipment, Doug the backup data recording, and I the fine guiding to keep Uranus and the star from drifting out of the view of our photometer. Mounted on the telescope, the photometer would separate the combined light from Uranus and the star into three different wavelength bands and convert each into an electrical signal that would be recorded on magnetic tape for later analysis. Signals from two of our photometric channels were also routed to a chart recorder that would give us a continuous graph of the detected light that we could monitor as the occultation progressed.

We expected that the occultation would go much as previously observed occultations involving Jupiter, Neptune, and Mars. The chart pen would first trace a constant level, with some jitter due to noise, that would represent the combined light intensity from the two objects. Then the signal would drop to the level of Uranus by itself, as the planet passed between the Earth and star. How the star flickered as the planet snuffed it out would tell us the Uranian atmospheric temperature; the precise duration of its disappearance would help establish the planet's exact size and shape. When the star reappeared from behind the planet, the signal would then rise back to its original value. Some of the light from the telescope would be diverted to a television monitor, so we could see the position of the image of Uranus. However, the telescope would smear the images of the star and planet into one big blur; from the television we could not tell when the occultation was occurring. The signal level on the chart recorder would provide our best information.

All seemed to be in order, but we had one major worry: there was a 17 percent chance that the northern edge of the shadow would pass south of us, meaning that we would not see an occultation! How could so many people go to so much trouble and expense to observe an event that might never happen? Well, it all seemed certain until about two months before. An error had cropped up in the prediction of the occultation, but by then plans had proceeded too far to be canceled for a small, though significant, uncertainty. As the time drew nearer, that 17 percent became larger in my mind. A miss would damage the reputation of occultation studies, which had just recently begun to yield respectable results for the atmospheres of Jupiter, Neptune, and Mars. How could one justify all this effort if the occultation did not occur? How could I justify watching Uranus pass near the star, but not actually occult it? I had no good answer to that question, only a farfetched one that my colleague Joe Veverka had hit upon in casual conversation in the hall one afternoon before we had left Cornell. He suggested that I use the data from a near miss to place an upper limit on a possible ring system of Uranus. This became a team joke, in that all of us knew—by habit—that Saturn was the only planet with rings, and we really did not think that anyone at NASA would appreciate an "upper limit on rings" as a justification for an ill-fated Southern Hemisphere expedition with the KAO.

We were not alone in our concern about a "nonevent." Our colleague from Lowell Observatory, Bob Millis, was set up to observe the occultation on a 0.6-meter telescope at the Perth Observatory. He too had accepted the possible miss as a risk worth taking in view of the reward for successful observations of the occultation. We had an important advantage over Millis, however: our airborne observatory would go as far southwest as possible in order to maximize our chances of being within the shadow of the occultation.

As the KAO swept through the turn onto the course for tracking Uranus, we made final preparations that were recorded on the intercom channel of the KAO. (--- indicates where the less relevant portions of the transcript have been left out.)

Elliot.	OK, we're in the turn.
Dunham.	Twenty hours, zero minutes, and zero seconds. [20:00:00 universal time (UT)]
Reisner.	Long way to go.

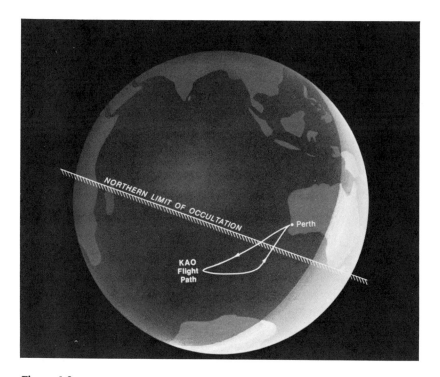

Figure 1.3
Round trip from Perth to Perth. This perspective of the Earth would have been
seen looking back from Uranus. At the right, the sunrise line was slowly creeping
westward, and the dashed line running across the globe marks the top of the
shadow of Uranus cast in the starlight of SAO 158687. Above the line, the star
never disappeared behind Uranus. Below the line, however, an occultation was
visible—seen only from Cape Town and the Kuiper Airborne Observatory, whose
round trip from Perth to Perth is shown in the figure. Not knowing the shadow
boundary before the event, we chose the flight path to be as far south as possible in order to maximize our chances of being within the shadow of Uranus. [Re-
printed, by permission, from James L. Elliot, Edward Dunham, and Robert L.
Millis, "Discovering the Rings of Uranus," *Sky and Telescope* 53 (1977), 413, copyright by Sky Publishing Corporation]

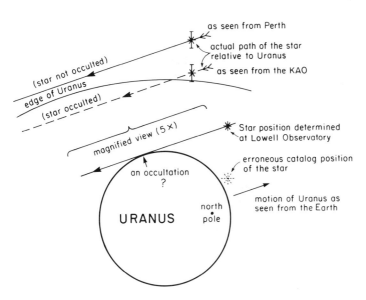

as seen from Perth

actual path of the star
relative to Uranus

as seen from the KAO

(star not occulted)

edge of Uranus

(star occulted)

magnified view (5X)

Star position determined
at Lowell Observatory

erroneous catalog position
of the star

an occultation
?

north
pole

motion of Uranus as
seen from the Earth

URANUS

Figure 1.4
A small error causes a big problem. This diagram shows the apparent motion of Uranus and the star as viewed from Earth. If the star had been located at its (erroneous) catalog position, an occultation would have been visible anywhere on Earth that Uranus could be seen above the horizon. However, new measurements placed the star considerably north of its catalog position, and the measurement error—indicated by the vertical bar—left the occurrence of the occultation in doubt. The odds of observing an occultation increased as one went further south because of parallax. The magnified view illustrates the effect of parallax for observing the occultation from Perth and the KAO.

We watch the wide field camera as the KAO nears the heading on which it can acquire Uranus.

Olson.	Down, getting closer.
Reisner.	It's going to be the left of the mount.
?	There's a light on the television.
Mink.	That's the Moon, off across the top.
Matheson.	[Earth] Satellite.
Reisner.	The Moon over Antarctica.
Oishi.	We'll see the Moon over Miami too.
?	Start the viewer.
Reisner.	On heading.
Mink.	The Moon left a nice afterimage there.

Meyer.	The object is centered. I have identified it.
Elliot.	You see it, Al?
Meyer.	Yeah, looking right at it.

Figure 1.5
Preparing for the moment of truth. Top: Telescope operator Milo Reisner (on the left), meteorologist Pete Kuhn, pilot Ron Gerdes, and mission director Carl Gillespie (back to camera) confer. Bottom: Seated in front of their data monitoring and recording equipment (from left to right) are Jim Elliot, Ted Dunham, and Doug Mink. [Courtesy of NASA]

Elliot.	OK, we can say it definitely hasn't occulted yet. The counting rates are right for the star and Uranus.

Continuous data recording begins at 20:05:40 universal time (UT), 41 minutes earlier than the predicted occultation time, to allow for error in the prediction.

Elliot.	OK, it took about nine minutes to get all set up and situated.
Dunham.	OK, the Cipher is still running properly.
Elliot.	ADAMS status?
Matheson.	Running—luckily.
Dunham.	Not yet. I want to get myself a little more organized.
Dunham.	Twenty hours, eight minutes, and forty seconds right there.

The Cipher is the digital tape recorder receiving our data, and ADAMS is the Airborne Data Acquisition and Management System, the computer on board the KAO.

Elliot.	Can you see on your tracker trace, Al, anything about the star?
Meyer.	No sign of asymmetry. We'll try a different scale.

Now we check our photometer signal levels for Uranus, star, and sky. While we are doing this, an unexpected dip in the signal goes unnoticed on the chart recorder.

Elliot.	We could do a little better check on the sky—you've got, what, six hundred for the sky?
Dunham.	I've got six hundred seventy, six hundred thirty, and one hundred seventy. Last time I got, let me look . . .
Elliot.	No, I'm talking about here. . . . I got the average ones.
Elliot.	There's about two thousand seven hundred, fourteen eighty, and four hundred, so, oh,
Dunham.	Wait a minute, that's photons.
Elliot.	No, these are the counts that we would get.
Dunham.	Well, that's odd because I wrote. . . . You're sure you don't have aperture three counts?
Elliot.	No, aperture two, twenty-seven seven seven; fourteen eighty and four-forty one.
Mink.	No.
Elliot.	[Garbled] for the sky.
Dunham.	Read those by again, more slowly.

Elliot.	For the sky, essentially, twenty eight hundred in channel one, fifteen hundred in channel two, four fifty in channel three.
Dunham.	OK, fine.
Elliot.	OK, now relate that to tonight.
Dunham.	We've got six seventy, six thirty, and one seventy.
Elliot.	OK, so you're down by two thousand counts, that should be twenty-four eight in the top one—OK, look'it there.
Dunham.	Pretty . . . good, pretty good. How about the bottom one?
Elliot.	Well, OK, if you want to do that one? We were down, . . . what was the sky tonight in the bottom one?
Dunham.	One seventy.
Elliot.	One seventy?
Dunham.	Yeah.
?	Well, subtract two hundred to get one thousand six hundred fifty . . .
Dunham.	What was that? What was that? [An unexpected dip in the signal has occurred on the chart.]
Elliot.	What?
Dunham.	This! [Pointing to the dip]
Elliot.	I dunno. Was there a tracker glitch?
Meyer.	Nothing here.
Mink.	Uh-oh.
Dunham.	No, I don't think it's anything here, it's clearly duplicated in both of those.
Elliot.	Yeah, I mean, clouds or . . . ?
Meyer.	Ask Pete.
Dunham.	Pete, what's your water vapor?
Kuhn.	Eight point nine.
Dunham.	Well, that's pretty low.
McClenahan.	What happened?
Elliot.	Well, we got a dip in the signal here which was either due to a loss or a momentary glitch in the tracker, or a cloud whipping through.
Dunham.	OK, I think someone should have the responsibility of always watching the focal plane there. I suppose a lot of people are.
Elliot.	But no one caught that one.
McClenahan.	Nobody caught that one. I didn't. I wasn't looking.

Gillespie.	There was no movement in the focal plane. I was watching it.
Matheson.	I was pretty much watching. I didn't see any.
Dunham.	OK.
Gillespie.	There's nothing the eye could detect, Tom.
Matheson.	. . . it was a pretty long period, too.
Elliot.	That, you should be able to see, because last night I saw a big glitch that went halfway out of aperture and back, and it was about like that on a plot we got.

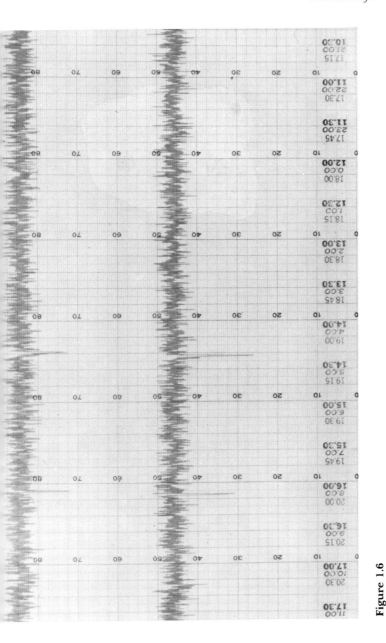

Figure 1.6

Odd flickers. The two traces on these graphs show the combined light of Uranus, star, and sky background plotted versus time, as we saw them traced by the chart recorder in our equipment rack. The raggedness of the traces is caused by the random arrival of photons in our detectors. The unexpected dips in the signal lasted a few seconds; we first thought they were caused by passing thin clouds or momentary failures in the telescope tracking system.

The unexpected dip bothered us, but remains unexplained.

Dunham.	Well, the long haul begins.
Many.	Yep.

Dunham.	OK, I got a deep short spike here. [A second dip]
Elliot.	I wonder if we're getting any clouds?
Dunham.	No, Pete said we had eight point nine microns of water.
Kuhn.	There's no clouds; I mean, truthfully, there's nothing up here.
Elliot.	Well, maybe this is a D ring. The D ring of Uranus.
Many.	[Laughter]
Mink.	Yeah.
Dunham.	With a normal optical depth of three, right?
Dunham.	Another one. [A third dip]
Mink.	Yep.

The possibility that these dips are not due to clouds or instrumental effects quickly becomes an exciting topic.

Elliot.	Yeah, those are real—I guess.
Oishi.	Boy, that was a deep one.
Kuhn.	Yep. There's no indication of any fog at all.
McClenahan.	Doesn't seem to be any bore sight shifting.
Elliot.	[Excitedly] Yeah, that's good. I think we're getting real—[garbled] could be small bodies—[garbled] the satellite plane is edge [face] on or it could be just small bodies like thin rings.

Uranus is tilted on its side so that satellite orbits would appear as a bull's-eye target.

McClenahan.	Are you centered up on the star now?
Elliot.	Yeah. There's no question about that. It was too quick for a tracking glitch, I think.
Mink.	They're too quick for much of anything.
Dunham.	You know what they might be; they might be a glitch with the star—how far away should the star be from Uranus right now?
Elliot.	It's close. It's really in close.
Dunham.	I was going to suggest that the star is going out of the aperture but Uranus not, but [Uranus and the star are] too close for that.

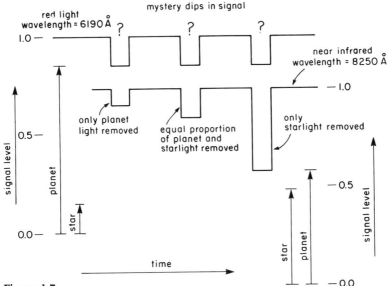

Figure 1.7
The color is the clue. Since the star and Uranus had different colors, the ratio of the dip amplitudes in different colors told us whether the light lost was only starlight, only Uranus light, or some combination of the two. From an approximate calculation we determined that only starlight was being lost, which led us to conclude that we had discovered a belt of small satellites around Uranus.

Here we realize that our separate channels, recording signals for different wavelengths of light, give us an important clue for determining the cause of the unexpected dips. Since the star and Uranus have vastly different colors, the ratio of planet light to starlight was not the same for our three photometric channels. By comparing the depth of the dips for the two channels on our chart recorder, we can determine whether only starlight is disappearing, only planet light, or some mixture of the two. Also, if only starlight were disappearing, then our chart pens should never drop below the level of Uranus alone, which we had established on a calibration flight 2 days earlier. In the following conversation, Ted and I develop these ideas.

Elliot.   And it's a pretty big fraction I mean . . .
Dunham.   Yeah, but remember if it goes all the way [full loss of signal from both objects], it's going to go to here.
Elliot.   Yeah, right, what I mean is that some of the fraction of the starlight itself.
Dunham.   Yeah.
Elliot.   Well, we can call and see if Millis is getting anything.
Dunham.   Yeah, that will be interesting.

Elliot.	The test is that they never should go below the intensity of Uranus. In fact, channel one is the real acid test of them, because Uranus is so bright.
Dunham.	That's right.
Elliot.	You see . . .
Dunham.	I know where Uranus is, Uranus is right there for channel three, but I don't know where it is for channel one.
Elliot.	Maybe it's something to do with Uranus, because they seem to be about the same amplitude on that scale. Well, I don't know, we're going to have to . . .
Elliot.	Nominal occultation in twenty minutes.
Mink.	Right.
Elliot.	Is that another one?
Dunham.	Another one! [A fourth dip]
Elliot.	Yeah, we'll see now, the relative amplitudes should be, I mean if they're bigger, channel two has the biggest.
Dunham.	This is a bigger span than this.
Elliot.	OK.

We now conclude that the amplitudes of the dips in signal at different wavelengths mean that only starlight is being removed.

Elliot.	It's definitely the star being occulted somehow.
Dunham.	Well—OK.
Elliot.	Yeah, the original thing being that this ratio to this is like that ratio to that.
Dunham.	Right, I agree.
Elliot.	Yeah, well, I told you every occultation comes up with a new surprising thing, and this may be it.
Dunham.	OK, it's twenty hours, twenty minutes, and twenty seconds.
Elliot.	OK, wait a minute, Millis wouldn't see the same ones, because if we're two thousand kilometers . . .
Dunham.	Yeah, but if you're seeing similar things . . .
Elliot.	Similar things, yeah . . .
Dunham.	How close are we to Sutherland? [Referring to the distance perpendicular to the motion of the planet's shadow across the Earth]
Elliot.	We're north of them.
Dunham.	Yeah, but we're a lot closer to them than we are to Millis, right?
Elliot.	Right, right.
Elliot.	They probably can't even see it yet, because it's down at ten degrees now, it's the preoccultation time.
Dunham.	There's another one! [A fifth dip]

Elliot.	No, you can tell, see, that all the channels, if we could look at them, they would have different relative ratios on each. Yeah, for Uranus going out, you'd get one signature; for the star going out, the other one. [This remark refers again to the different colors of Uranus and the star.]
Dunham.	Channel one will actually be an acid test, won't it?
Elliot.	Yes, I'm very glad we have it.
Elliot.	Yeah, we're always joking about this D ring; it looks like maybe we've got one.
Dunham.	The watched pot never boils, here, we haven't got one so far.
Elliot.	That's OK, we've got enough already.

[Everyone talking at once]

Elliot.	If you get them all the time, it's going to be bright enough to see the darned thing as a ring.

The combined light from many small bodies around Uranus would be bright enough to have been detected in photographs of Uranus. Finding too many small satellites would then conflict with the fact that Uranus was known not to have a visible ring.

Elliot.	What'll we call them, what are we going to have as a name for them? Let's see, think of something classic.
Mink.	Uranus has its own asteroid belt, obviously.
Reisner.	They're retracks.
Many.	Retracks, yeah! [Laughter]

We think we have discovered a "belt of satellites" around Uranus. If so, we can confirm this discovery by extending our planned observing leg to pick up satellite occultations from the portion of the "belt" on the other side of Uranus.

Elliot.	Well, with all this stuff, we can't leave the leg early, we'll run it out to the end, because they're gonna be coming on the other side.

Dunham.	I don't know, maybe this [space that we are probing now] is in between the ring and the planet.
Elliot.	Yeah, it has to be some kind of stable thing, I hope it is.
Mink.	It would be nice if we got them symmetrically on the other side.

Gillespie.	I know it will have to be checked with the altitude, but probably there are clouds of orange juice floating around up there.

[Laughter]

McClenahan.	Sure sign of life.
Many.	That's right.
Dunham.	Junk in orbit.
Elliot.	Old satellites that were launched by an ancient race of Uranians.
Gillespie.	They took to launching orange juice cans in orbit.
Reisner.	Swarm of honeybees looking for a place to settle, between point four six and . . .
Dunham.	It's been a long time before we really got a good one.
Elliot.	Yeah, it must be some sort of, I mean if there really are things in orbit, there must be some sort of belt of them.

Elliot.	We never could have thought of it. All these occultations are great, you never think of what's going to come on the next one, see . . .
Gillespie.	That's what makes them interesting.
Dunham.	I remember . . .
Elliot.	I told Ted I thought on this one we've run out, there will be nothing new happening on this one, because we'll have done them all.

We now wait for Uranus to occult the star.

Elliot.	It's going to be late, if it is.
Dunham.	Just on the general principle that it can't be on time.
Mink.	Right!
Mink.	It's obviously not going to be very early, anyway.
Dunham.	Twenty hours, forty-seven minutes, and two seconds.
Elliot.	Yeah.
Olson.	The pilot's doing a good job.
Matheson.	Yeah, the pilots are going real smooth.

Elliot.	We're five minutes late, now we're up to the Perth duration.

The occultation is late, probably indicating that the shadow of Uranus is further south than predicted; but how far south?

The signal becomes noisier and rapid variations or "spikes" begin to appear.

Dunham.	These things are starting to look real.
Elliot.	Yeah.
Mink.	Yeah.
Elliot.	We might be starting to occult.
Dunham.	Yep.
Kuhn.	No clouds at all, just crystal clear.
Dunham.	It's right in the middle of the aperture, yeah, OK, we're starting to get in on it here.
Dunham.	The cassette will last until this is over. Yeah, without a doubt, this is the thing.
Elliot.	Well, I don't know.
Dunham.	OK, I'm starting the burst mode on this—they'll run for a hundred seconds.
Dunham.	Yeah, Elliot, that's definitely it.
Elliot.	Ahhhhhh . . . !!

[Loud whoopees, yays, etc.]

Dunham.	Oh, look at the spikes. Boy have we got a lot of those.
?	Hey, disappearance.
?	There you go. Boy, you've got more than enough.
McClenahan.	Another Ph.D. in the world.

Elliot.	We have to really give Wasserman credit, boy, he really called this one to a few minutes. But that's more spikes, more than Jupiter, I think, almost.

Elliot.	Did you tell the pilots everything was going great back here?
McClenahan.	Yeah.
Elliot.	Good, OK. Tell them to stick with it, and we got to take the leg until we get wiped out by dawn, because of those blips at the beginning. I think we might pick them up on the other side.
Gillespie.	I told them before we took off that they'd have to go all the way to dawn.
Elliot.	OK.

[Laughter]

Meyer.	I've never seen an experiment quit early yet.
Matheson.	And he'll ask for an extension, right?

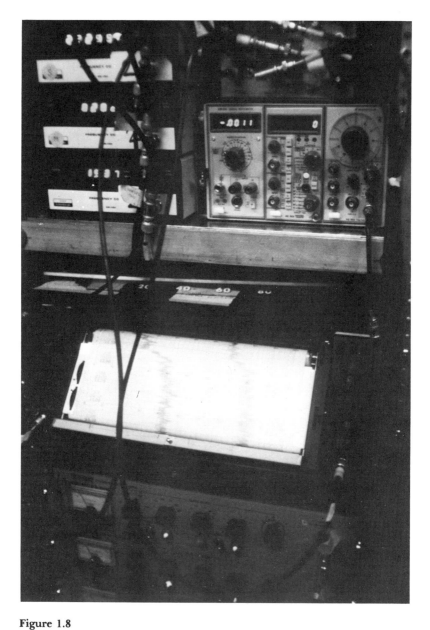

Figure 1.8
The star disappears behind Uranus at last! About 5 minutes after the predicted time, Uranus occulted the star. The chart record shows numerous "spikes," abrupt flashes of light caused by density variations in the Uranian atmosphere of only a few percent. [Courtesy of Doug Mink]

Elliot.	On to Pago? [Samoa]

Dunham.	The sun is ten degrees down at twenty-two hundred, right?
Elliot.	Yeah, we'll start picking up sky [brightness] a little before that, but . . .
Elliot.	. . . but those other things [dips] were fast, so we'll be able to just interpolate the baselines.
Dunham.	Yeah, sure.
Dunham.	Yeah, this is really very good. I wonder how many of those blips he [Millis] got?
Elliot.	I don't know, he wasn't going to start his tape until right before [the predicted occultation by Uranus], so he probably has them on his chart and probably thinks they're tracking problems.
Dunham.	Yeah.
Mink.	Maybe we can compare times.
Dunham.	For whatever that's worth.
Dunham.	Except he's two thousand kilometers away.
Elliot.	. . . time scale of [those] other things [dips], a second or two? I mean like the star—that's about eleven kilometer objects.
Dunham.	Yeah.
Elliot.	We may even be able to see some diffraction from them too.
Mink.	Oh, wow! I can run my Fresnel patterns on them.
Elliot.	They're irregular things so they would be . . . sort of funny.

Mink.	So that means we'll have twenty-two minutes duration.
Elliot.	Oh-oh. Millis is close. Because [the duration of] his [occultation] was supposed to have been twenty-two minutes itself.
Dunham.	Yeah.
Elliot.	I think we went about as far south [from Perth] as he was [predicted to be] from the northern edge [of the shadow].

From the duration of the occultation observed from the KAO, we realized that the northern edge of the shadow of Uranus may have passed south of Perth. We then roughly figure out when we should start seeing the dips again, which should also occur when the star would pass behind the "satellite belt" on the other side of Uranus.

Elliot.	Run out the leg. What's the midtime of the earlier . . . so we can see where we should start picking up those negative ones.
Mink.	Twenty-one oh five, twenty-one oh six about.

The dips reappear, as expected. No one is surprised. We would have been surprised if the dips didn't reappear.

Dunham.	OK. We got one. We got a blip here.
Elliot.	All right a blip. OK. Here we go.

Elliot.	We got blips again.
Mink.	What was the time about?
Dunham.	Maybe a half-minute before that.

Mink.	That was one at 21:47.
Elliot.	They're real.
?	Yeah.

As the dips began to appear again, I was composing a note to alert South African observers to the "belt of satellites" that we had discovered and to suggest that they continue their observations as long as possible, since dawn would force us to stop recording data several hours before it would force them to quit. Who knew how far from Uranus the "belt" extended?

Although the term "ring" appeared many times in our conversation, we thought we had found a "belt of satellites" that surrounded Uranus. If this belt contained enough satellites, they would collectively reflect enough light to be photographed from Earth and would appear as a broad ring. Since a photograph would have a resolution 10,000 times worse than we had achieved with the occultation data, it could not resolve the individual bodies. At this point, narrow rings were not considered a serious possibility. We knew only of Saturn's broad rings, which were tens of thousands of kilometers wide in the radial direction. I had brought up "thin rings" in conversation, but serious consideration of narrow rings—only 10 kilometers wide—was too big a leap for our imagination at that stage. Nevertheless, we knew that we had discovered something exciting—it turned out to be the first discovery of a planetary ring system in more than 350 years.

2

The First Rings

The detection of the rings of Uranus was not the first time that recognition, the crucial second step of discovery, had lagged behind observation. In the early seventeenth century, astronomers confronted a similar recognition problem for more than 50 years before seeing the truth. As soon as Galileo Galilei turned his tiny, crude, spyglasslike telescope on Saturn in 1610, he knew there was something odd about the farthest known planet. Only a bit more powerful than a pair of field glasses and full of distortion, his first telescope revealed to Galileo's eye that "the planet Saturn is not one alone, but is composed of three, which almost touch one another . . . and the middle one is about three times the size of the lateral ones. . . ." To preserve his rightful credit for the discovery of Saturn's companions while he worked on a formal publication, Galileo released a cryptic message, an anagram, whose solution was "I have observed the highest planet to be triple bodied."

The credit for his discovery was secure, but Galileo would never decipher the true nature of his find. And an odd nature it must have seemed. Having discovered the four starlike companions of Jupiter only a few months before, he assumed that these two larger bodies would also revolve about their planet, but as the months passed they refused to budge. When he happened to resume his casual observation of the planet in 1612, they had disappeared. "Has Saturn, perhaps, devoured his own children?" he asked. As he expected, they did return, but the surprises were not over. When Galileo looked again in 1616, Saturn was in a new guise. Through the same telescope, Galileo saw "two half-ellipses with two little dark triangles in the middle of the figures and contiguous to the middle globe of Saturn. . . ." Squint just a bit at a modern photograph of Saturn and you will see Saturn as Galileo described it and drew it in 1616. The difference is that we know what we are seeing. Galileo's eye saw, it appears, but his mind did not.

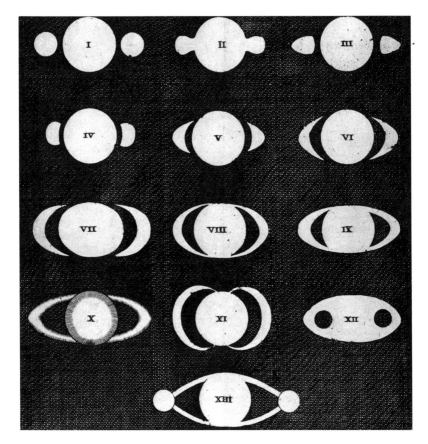

Figure 2.1

Early rings. These drawings from Huygens's *Systema Saturnium* of 1659 illustrate the variety of perceptions of Saturn that resulted from the interplay of its changing orientation, improvements in telescopes, and new interpretations of its physical nature. The observers were I, Galileo (1610), who in 1616 drew Saturn much like IX; II, Scheiner (1614); III, Riccioli (1641 or 1643); IV–VII, Hevel (theoretical forms); VIII and IX, Riccioli (1648–1650); X, Divini (1646–1648); XI, Fontana (1636); XII, Biancani (1616); Gassendi (1638, 1639); XIII, Fontana and others at Rome (1644, 1645). [Reprinted from A. F. O'D. Alexander, *The Planet Saturn: A History of Observation, Theory and Discovery* (London: Faber and Faber, 1962), facing p. 96]

In the 1640s and early 1650s telescopes magnified more and distorted less, but even frequent observation did not reveal the true nature of the phenomenon. The best drawings did show an apparition that emerged from invisibility on either side of the planet, gradually grew to rival the planet itself, and then shrank back to invisibility, all between 1642 and 1656. The changing appearance of Saturn did not remain mysterious for lack of theorizing. Astronomers offered all manner of geometrical contrivances to explain the planet's behavior—two crescents attached to a tumbling planet; two large, dark satellites (the dark triangles) and two bright ones outside of those; or, perhaps an egg-shaped planet with four black spots. The most popular idea was some sort of vaporous exhalation of the planet condensed into an ever changing cloud or a thin elliptical corona that rotated with the planet.

The Dutch astronomer Christiaan Huygens eventually solved the puzzle, but not necessarily because he had a sharper view of Saturn. Instead, he made some crucial observations and combined them with a working theory of how satellites behave. In March 1655, he discovered Titan, Saturn's largest moon, circling the planet every 16 days. By this observation and by noting the path of Saturn against the background of stars, Huygens determined that the ansae, or handlelike protuberances of the planet, as well as the orbit of the satellite were tilted more than 20 degrees with respect to the plane of Saturn's orbit, that is, the ecliptic; Saturn, he saw, was tilted much as Earth is. He could also see that the ansae did not shrink toward the planet as they winked out during the winter of 1655–1656.

These astute observations meshed nicely with Huygens's belief in Descartes's theory of planetary vortices. By analogy with the difference between the 1-day rotation of Earth and the 27-day revolution of Earth's moon, everything inside the orbit of Titan, including the ansae, should revolve around Saturn in less than Titan's 16-day period of revolution, Huygens reasoned. That rotation would be about the axis perpendicular to the plane of Titan's orbit, the plane that also contained the line of the ansae. Because Huygens could see no change in the appearance of the ansae in a period as short as 16 days, he concluded that the rotation of the ansae must be symmetrical about the axis. This geometrical constraint eliminated a lot of possibilities. A flat, wide ring, he realized, could rotate symmetrically while appearing unchanged. Such a rotating, inclined ring would also remain fixed with respect to the stars but show first one side then the other to the Sun and Earth-based observers. In between, a ring would be edge on and invisible, as observed. From the perspective of the Sun, the Earth performs the same apparent wobble as it progresses about its orbit—

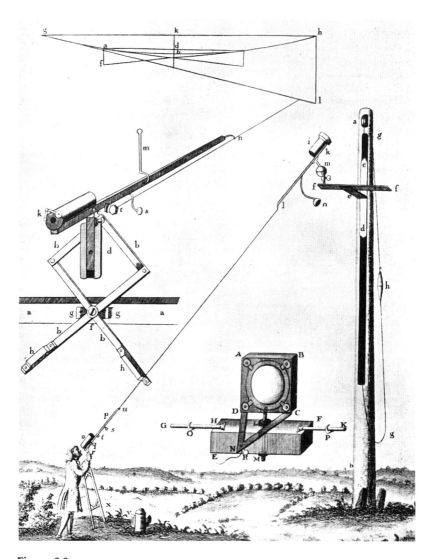

Figure 2.2
Huygens at his telescope. [Reprinted from Henry C. King, *The History of the Telescope* (Cambridge, MA: Sky Publishing Corporation, 1955), p. 55, copyright by British Crown]

Figure 2.3
Cassini splits the ring. In 1675 Cassini reported that a "dark line" divided the
ring into two parts, "the inner of which was brighter than the outer one." He
made this drawing in 1676, apparently showing the outer ring to be slightly narrower than the inner ring. [Reprinted from A. F. O'D. Alexander, *The Planet Saturn: A History of Observation, Theory and Discovery* (London: Faber and Faber, 1962),
p. 115]

the Sun beams down from above the equatorial plane (summer), moves
to stand over the equator (spring or fall equinox), and then shines
from below the equatorial plane (winter).

Huygens announced his discovery of Titan in a pamphlet in March
1656 and included his solution of the Saturn problem as an anagram;
he needed time to tie up a few loose ends. Even knowing of Titan's
existence, no one proposed anything like Huygens's "thin flat ring,
nowhere touching, and inclined to the ecliptic" before his public announcement in 1659. Those who did not immediately defer to Huygens's simple logic acquiesced in the face of his counterarguments
and his increasingly accurate predictions of Saturn's future behavior.

Even as observers began rendering Saturn as a ringed planet, the
way theorists told them they should, Jean Dominique Cassini announced
in 1675 the discovery of a dark line that seemed to divide the ring
in two. During the next 300 years, Cassini's division would be the
only ring feature that everyone could agree actually exists. As it would
centuries later, the division raised questions among physicists of the
day, according to the historian Albert Van Helden. It aggravated an
existing problem they had with the solid, relatively thick ring espoused
by Huygens. His ring disappeared when seen edge on because of a
dark, absorbent edge or perhaps because of reflection from a single
point on a curved, mirrorlike edge. The simplest explanation, his critics
argued, would be that the ring was too thin to detect edge on. Simpler
might be better, but a thin ring, even in the days before Newton
offered his new dynamics, seemed too fragile to stand up under the
strain of rotating so close to Saturn. Cassini's division only made things
worse.

As early as 1660, a poet, Jean Chapelain, suggested a solution to the thick/thin dilemma—perhaps the solidity of the ring was only an illusion. "A swarm of small satellites," in Cassini's words, would give the appearance of solidity when viewed from so far away as Earth, but would require no new dynamical explanations. Each particle in the ring would orbit the planet just as the Moon orbited Earth, Galileo's moons orbited Jupiter, and Titan orbited Saturn. Cassini's division would make no difference at all. According to Van Helden, the idea of particulate rings gained favor toward the end of the seventeenth century, becoming the consensus before its close.

Then, something curious happened. Science took a step backward. William Herschel, the preeminent astronomer of the eighteenth century, pronounced the ring to be solid, and so it became. He had read it in the writings of Huygens, who died believing it. Besides, it made sense to Herschel. He resisted the division of his solid ring by Cassini's feature, preferring to see it as a marking of only one side of the ring. He would have to see a star in the supposed gap during its passage behind the ring, he said, before he would believe that there was empty space dividing the rings. Eventually, Herschel settled for seeing identical markings in the same position on both sides of the ring. There might be two rings, Herschel conceded, but they were still solid.

Herschel had the help of an influential theorist, Pierre Simon de Laplace, in his revival of solid rings. To Laplace, the rings of Saturn were a window into the early days of the solar system when matter progressively condensed to form planets and satellites. The rings were an arrested, petrified stage in his nebular formation of the solar system, during which the ball of material that would become Saturn was shedding rings that would coalesce into solid satellites. The present rings of Saturn never broke up to form satellites, but solidified as is. And a great many rings there must be, Laplace concluded. Unlike Herschel, he found a single ring or even two rings to be impossibly unstable—only at a single point across the width of a rotating ring would the outward push of centrifugal force balance the inward pull of gravity. The imbalance of forces elsewhere on a wide ring would tear it apart no matter how thick it was, Laplace said. His solution was to subdivide the two known rings into a myriad of narrow rings nested inside one another, thus reducing the strain across any one ring. By concentrating the mass of each ring off center, he further reduced the strain. What he actually did was make a ring behave as much like an orbiting particle as possible while remaining a solid ring. Still, he would not abandon solid rings entirely, committing astronomers to them for the next 50 years.

Laplace said that theory demanded a multitude of ringlets, and observers did not let him down. Ring subdivisions during the first half of the nineteenth century multiplied into a bewildering array that seemed to back Laplace's view of circles within circles, each sliding by its neighbors as it spun about the planet at its own pace. On an evening in 1825, the Englishman Henry Kater saw three divisions in the outer ring alone, but reported that there might be more. A friend viewing through the same modest 7-inch reflecting telescope that evening reported six divisions in the same ring. The German Johann Encke saw only a single line or stripe as broad as Cassini's division in the middle of the outer ring in 1837; it could have been the central marking of Kater's three. The inner ring at times appeared featureless, at others divided into two, three, or more rings. Charles Tuttle and two assistants reported in 1851 that their 6 1/3-inch refractor showed the inner ring to be "minutely subdivided into a great number of narrow rings . . . not unlike a series of waves."

Mid-nineteenth-century astronomers did not confine their predilection for finding new rings to Saturn. Within 2 weeks of the discovery of Neptune in 1846, the noted English amateur astronomer William Lassell wrote *The Times* to convey the news that he "suspected the existence of a ring round the planet." He also announced the discovery of a satellite. Other observers confirmed both the satellite and the ring, the latter appearing variously as an oblong appearance of the planet, a distinct ring, or "a cluster of satellites." One astronomer even reported a tilt of the ring that was within 4 degrees of Lassell's estimate.

Still, Lassell, of all people, remained unconvinced. Greater astronomers had been misled, he knew. William Herschel had also had suspicions of a ring or rings about the newly discovered Uranus. Ever cautious that his apparent ring might be an optical aberration, Herschel eventually gave it up as a deception. Richard Baum has recently attributed Herschel's phantom ring to a classic case of astigmatism—in Herschel's telescope, not his eyes. Lassell too may have had the same problem, Baum concludes; however, Lassell's colleagues' ready support cannot be so easily explained away. In any case, the purported rings of Neptune eventually faded from sight.

The Neptune affair did not stop astronomers from reporting a profusion of rings around Saturn. The large number of rings fit Laplace's theory, but their variability in number and location cast doubt on their solidity. Only Cassini's division seemed anchored and unchanging. The seeming mutability of the rings drove George Bond to propose in 1850 that the rings must not be solid or rigid, but somehow fluid. His reevaluation of Laplace's calculations only reinforced his feeling that

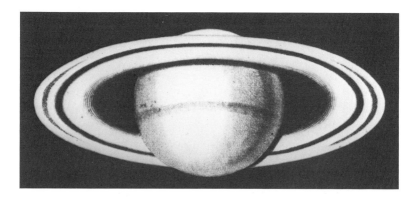

Figure 2.4
Two features of the outer ring. In 1837, J. F. Encke (top) used a 22.5-centimeter telescope to detect the feature shown in his drawing. He described it as a broad, low contrast line or stripe in the middle of the outer ring. [Photograph courtesy of D. Osterbrock from D. Osterbrock and D. Cruikshank, "J. E. Keeler's Discovery of a Gap in the Outer Part of the A Ring," *Icarus* 53 (1983), 165–173] In 1888, James Keeler (bottom) detected a thin, dark line, an apparent gap, about one-fifth of the way in from the outer edge of A and at the outer edge of a broad feature. The narrow gap is now called the Encke division. [Yerkes Observatory photograph courtesy of D. Osterbrock]

Figure 2.5
Finally, the third classical ring of Saturn. In November 1850 William Bond and his son George saw a dark band across the planet immediately adjacent to the inner edge of the inner ring. They did not recognize it as a ring for several days until Charles Tuttle, their assistant, suggested that it could be caused by a dusky ring. Only 2 weeks later, William Dawes independently came to the same conclusion. The timing is more remarkable because the dark band had been noted and recorded by various observers for centuries without being recognized as a ring. The new inner ring has been called the crepe ring or the C ring. [Reprinted from A. F. O'D. Alexander, *The Planet Saturn: A History of Observation, Theory and Discovery* (London: Faber and Faber, 1962), facing p. 208]

solid rings would not work. The discovery of a third major ring in the same year by Bond and his father William, and independently by William Dawes, would prompt the final rejection of solid rings. This faint new ring was interior to the others, and was designated ring C to distinguish it from the middle, or B, ring and the outer, or A, ring. Not only was the C ring the faintest by far, but in 1852 several observers realized that they could see the planet right through the new ring. That was hardly Laplacian.

Explanations were not long in coming from theoreticians. First, James Clerk Maxwell won Cambridge University's Adams Prize in 1857 by demonstrating to everyone's satisfaction that Newtonian physics would not tolerate solid rings, no matter how finely subdivided. The irregularities demanded by Laplace would be so huge as to be visible from Earth, Maxwell said; even when stabilized by such mountainous terrain, the slightest disturbance would shatter the rings. Liquid rings, on the other hand, would break up and form visible satellites. Echoing Chapelain and Cassini of 180 years earlier, Maxwell reached the "new" conclusion that "the only system of rings which can exist is one composed of an indefinite number of unconnected particles revolving around the planet with different velocities according to their respective distances." Maxwell, unlike his predecessors, had marshaled irrefutable physical arguments requiring particulate rings, or so he thought. Mid-twentieth-century physicists found his upper limit on the density of stable liquid rings to be much too low—theory could not eliminate liquid rings after all. Maxwell's contemporaries did not catch the error, though, so they immediately enshrined particulate rings as conventional wisdom. No harm done. No one ever reported the image of Saturn glimmering off the rings, as would be expected of liquid rings. The rings did indeed reflect light as a swarm of small satellites should. And in 1895 James Keeler used his new spectroscope to show that the orbital velocity of the ring material varied across the rings just as would be expected if they were composed of tiny satellites—the inner parts moved faster than the outer parts. Stronger proof would come later.

If all of those reported subdivisions in the rings were not the boundaries between Laplace's wheels within spinning wheels, what were they? All illusion? The American astronomer Daniel Kirkwood had an answer—Cassini's division and Encke's division, at least, resulted from gravitational effects of satellites of Saturn. That was a start. Others would expand on that explanation for the next 100 years. The more conservative Kirkwood began his explanation in 1866 with the asteroid belt, a sparsely populated ring of rocks orbiting the Sun. According to his theory, Jupiter, the giant planet just outside the asteroid belt, clears the gaps seen in the belt through the long reach of its powerful gravity. Being closer to the Sun than Jupiter, each asteroid moves faster along a shorter orbit, overtaking and passing the plodding, vastly more massive planet after more than one of the asteroid's own revolutions. The timing of these close passages is crucial to Kirkwood's gap-clearing mechanism. If the asteroid's orbital speed brings it abreast of Jupiter at the same point or points in its orbit time after time, the effect of the stronger pull of the planet's gravity at these closest approaches could accumulate, nudging the asteroid into a new, more stable orbit. The timing of the nudge is like the timing of the push of a swing—the swing will not go higher unless the pushes come at the right place each time. An asteroid whose period of revolution about the Sun is one-half that of Jupiter, for example, would catch up with the planet at the same point after two complete revolutions of its own. Kirkwood argued that such orbital commensurabilities or resonances must clear the conspicuous gaps in the asteroid belt because the gaps appeared where asteroids would have orbital periods 1/2,

1/3, 2/5, and 3/5, that of Jupiter's. The match between the locations of resonances and gaps seemed impressive.

In 1867, Kirkwood extended his resonance theory to the case of Cassini's division. A ring particle orbiting within the division, he noted, would have an orbital period equal to half that of the satellite Mimas, about a third that of Enceladus, a quarter that of Tethys, and a sixth that of Dione. The 2 : 1 resonance of Mimas, in particular, because of the satellite's proximity to the rings, would make particle orbits in that division too unstable; resonances must keep Cassini clear. In 1871, Kirkwood included Encke's division among resonance-cleared features, although he thought it probable that Cassini's was the only permanent, completely empty gap. The rest of the subdivisions were evanescent; perhaps they were the shifting waves of darker shading that seemed to have so confused observers of the past half-century. Instead of the gaps between Laplace's solid rings, astronomers now expected to see subtle shadings where the gravity of Saturn's satellites had thinned out but not totally removed the ring particles, the way a light breeze might thin a mist.

The last decade or so of the nineteenth century saw a particularly severe "outbreak of division finding," as Saturn's chronicler A. F. O'D. Alexander put it, sparked in part by Kirkwood's theorizing and in part by vivid reports of ring features made by E. M. Antoniadi. Encke's division at times showed up somewhere near the middle of the A ring; the B ring often had several subdivisions, but could be free of markings at other times; some observers saw a division between B and C, while others swore there was none there; and ring C at times had markings and at other times not. Most of these features coincided with resonances, conveniently enough. There were even reports of white spots and radial markings on the rings. As Alexander observed from the perspective of 1962, " . . . skeptics might be inclined to dismiss them all as due to optical and atmospheric causes, differences between depth of shading in adjoining ring areas, or even the bias of preconceived ideas: the difficulty is to decide which of these temporary appearances were truly objective and which were not."

The problem of knowing whose eye at which telescope on which night could be trusted to see the true appearance of the rings persisted for 80 years until the objective eyes of spacecraft took a close look. The decline of the art of planetary observation only aggravated the problem. Distant galaxies lured astronomers away from the demanding discipline of planetary observation even as the debacle of the Martian "canals" discredited the observer's eye and drawing hand as recorders of delicate planetary detail. If the camera could not record it, why should anyone believe it was there? Percival Lowell had not only seen distinct "canals" on Mars, he also glimpsed dark patches and strips beneath the impenetrable clouds of Venus. As might be expected, Lowell found delicate details in the rings of Saturn, too; of course, most coincided with one resonance or another.

Planetary studies were in the doldrums during the first 60 years of the twentieth century, but those times still produced some constructive observations and insightful theorizing. In 1920, Sir Harold Jeffreys went beyond the best observational evidence and concluded that the rings of Saturn were vanishingly thin. True, after Huygens and his predilection for a thick ring, no estimate of the thickness of the rings had implied that they were anything but extremely thin. William Herschel inferred an upper limit of 450 kilometers by comparison of the edge-on rings with the diameter of the satellite Rhea. At first blush, 450 kilometers may seem thick, but the rings' diameter spans 274,000 kilometers. A 274,000-kilometer pancake would tower over such rings. A phonograph record of the same breadth would be five times thicker. Still, estimates plummeted from Herschel's 450 kilometers to George Bond's limit of 67 kilometers to a 1919 limit of 15 kilometers, a thickness like that of a stadium-size sheet of cardboard.

Jeffreys, without ever looking through a telescope, predicted that astronomers would never measure the thickness of the rings—theory required that they be too thin to detect edge on. To begin with, collisions between particles are inevitable, Jeffreys noted. Collision robs colliding particles of a bit of their energy of motion, and that kinetic energy is what would, in the absence of collisions, keep particles from forming an immeasurably thin ring. The stolen energy goes into deforming and heating the particles, processes which take up kinetic energy but cannot return it. Even if they began by swarming about the planet in every possible orbit, particles constantly losing energy through collisions would collapse into the planet's equatorial plane. There they would be in the lowest energy orbits, near the equatorial bulge of the planet. Once a ring formed, Jeffreys reasoned, collisions would continue, grinding much of the ring particles into powder and shaping the orbit of every chunk and mote into a circle. In the end, each particle would lie in the same plane as all the others as it continued to jostle gently the particles slightly inside and outside its own orbit. Jeffreys imagined that such a ring might well separate into circular strips of closely packed particles that allowed only a trickle of light to pass through them. To Jeffreys, Saturn's rings were a series of thin but dense dust piles.

However tightly packed Jeffreys's rings might be, observers were finding that they could, when chance offered the opportunity, see right through the bright A and B rings. It was a chance discovery despite the suggestion made 150 years earlier by William Herschel that the true nature of Cassini's division might be determined by waiting for a star to pass behind the rings and shine through the supposed gap. E. L. Trouvelot had suggested in the late nineteenth century that such a stellar occultation could also show whether the rings were transparent. No one ever took up the idea, but in 1917 it forced itself upon astronomers. Maurice Ainslie and John Knight, both British amateur astronomers, happened to be observing Saturn independently on the evening of 9 February when each noticed that Saturn's motion against the stars would soon carry the rings in front of a seventh-magnitude star. It shone brilliantly through the apparently empty Cassini division. It even managed to glimmer through the A ring, if only at 25 percent of its full brightness. Ainslie reported that the star showed "little or no variation in brightness . . . " except twice when it doubled in brightness near the outer edge of A. He assumed that one of the abrupt brightenings about one-fifth of the way in from the edge was the star breaking through the Encke division and that the other, even nearer the edge, was another gap.

Stellar occultations looked promising. Being a vanishingly small point of light, a star could probe the rings for gaps or thin spots that are far smaller than could be resolved by any telescope, no matter how good the atmospheric viewing conditions might be. Astronomers did not leave the next occultation opportunity to chance. Predicted 3 months ahead, bad weather wiped out all the planned observations in 1920 except at Rondebosch, near Cape Town, South Africa. This time, the star showed through the nearly edge-on B ring, the brightest and presumably densest of the rings. The star's "light fluctuated considerably and once gave a momentary flicker."

Although intriguing, the flickerings and fluctuations of occulted stars could never make much sense unless someone routinely predicted occultations and many observers studied them. They did not. According to Alexander, ring occultations in 1939, 1957, and 1960 attracted single observers to each. J. E. Westfall observed the 1957 event from California, reporting "many rapid rises and falls in brightness." The star shone brightly through the Cassini division and one or two apparent divisions near the outer edge of the A ring. Westfall also reported a slight increase in transparency in the central A ring. The exceedingly faint star of the 1960 occultation likewise disappeared and reappeared several times. Some reasons for neglect are clear. Occultations of bright

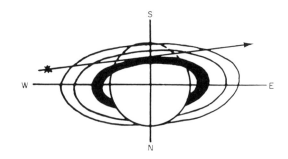

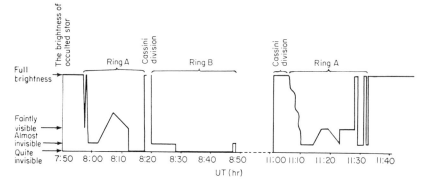

Figure 2.6

Westfall's stellar occultation observation. On 28 April 1957 J. E. Westfall observed the 3.5-hour occultation of a star by the A and B rings. The upper diagram shows the path of the star behind the rings. The lower diagram, by Bobrov, is a schematic representation of Westfall's written report of his visual observations. Peaks of full brightness, as in the Cassini division, equate with dark, empty gaps in drawings of the rings. The less visible the star, the brighter that area would be in a ring drawing. Notice in particular the apparent gaps near the outer edge of the A ring, the inner one being near Keeler's reported gap. There is also a broader, less distinct feature in the middle of the ring that is reminiscent of Encke's reported feature (figure 2.4). [Reprinted, by permission, from M. S. Bobrov, "Physical Properties of Saturn's Rings," in *Surfaces and Interiors of Planets and Satellites,* edited by A. Dolfus (New York: Academic Press, 1970), p. 399, copyright 1970 by Academic Press] stars are rare. If predicted, astronomers could not observe them with their sensitive light-detecting instruments because of the glare of the rings' reflected light; the imprecise and fallible eye would have to suffice. Photometric detectors had better targets in the rings themselves, where they would find increasing application to the problems of particle size, shape, and spatial distribution. For whatever the reason, there was little interest in occultations.

Visual observations of any kind may not have been popular, but one man at least kept his eye on the rings. In the 1940s, near the end of his life, Bernard Lyot devoted much of his time at the 24-inch refractor of the Pic-du-Midi Observatory high in the French Pyrennees to drawing the rings as he saw them. His rings were formed by a dozen or so subdivisions of the three major rings, consisting mostly of fuzzy "minima and maxima" of light. He did draw a narrow, black line about a fifth of the way in from the outer edge of the A ring, another distinct gap between B and C, and a pair of gaps near the inside edge of B. The feature in the outer part of ring A bore a striking resemblance to a fine, black line drawn by James Keeler in 1888 on the inaugural night of Lick Observatory's 36-inch refractor. Although Keeler's drawing from that night was widely circulated at the time, planetary observers seem to have forgotten it over the years. Lyot also drew Encke's division, rendering it as a broad zone of darker shading in the middle of A subdivided by three brightness minima.

Lyot's boldly shaded drawing, published in 1953 after his death, caused little stir in the astronomical community. Not that there were many bona fide members of the community of planetary observers. Gerard Kuiper, the dominant figure in the field, did take an interest in Lyot's conclusions. As the Herschel of his time, his word carried considerable weight. On a night of excellent viewing conditions, when he expected to see the finest detail possible, Kuiper peered through the 200-inch reflector on Mount Palomar, the mightiest telescope in the world. He did not see much. "*Only one division exists* [italics in original], the Cassini division, whose width is one-fifth of that of Ring A. The other 'divisions' are either minor intensity ripples, with some 10–15% amplitude, or are non-existent. The Encke 'division' is a ripple where at the same time Ring A changes its intensity abruptly." To Alexander, at least, the lesson was clear: "The importance of this observation and pronouncement by one of the world's leading planetary observers, using the world's largest telescope, is self-evident, and clearly refutes any idea that may exist of there being any gap through the rings other than Cassini's division."

Figure 2.7
Lyot's view of the rings. This drawing, whose contrast was somewhat enhanced, is Lyot's 1943 summary of his observations of the rings. Interesting features include (from the outside in) a narrow gap near the outer edge of the A ring, where Keeler reported a similar gap (figure 2.4); the broad central darkening of ring A like that reported by Encke (figure 2.4); three bright subdivisions of B; and a darker subdivision on the inner edge of B. [Reprinted from A. F. O'D. Alexander, *The Planet Saturn: A History of Observation, Theory and Discovery* (London: Faber and Faber, 1962), facing p. 400]

Kuiper failed to convince at least one observer of the blandness of Saturn's rings. That was Lyot's assistant, Audouin Dollfus. Dollfus asked and received permission from Kuiper to use in 1957 the 82-inch McDonald reflector, Kuiper's home telescope. Dollfus saw the same rings as Lyot had, gaps and all. His and Lyot's observations, usually in the form of seemingly more objective plots of brightness versus radial position, usually received some attention in reviews of ring studies. Some researchers took them seriously, some did not. Their drawings most impressed those who had been lucky enough to have exceptionally good views of the rings. In a 1975 letter to the popular astronomy magazine *Sky and Telescope*, W. C. Livingston told how he and George Abell, as "a social prelude to the night's work," had turned the 100-inch Hooker telescope at Mount Wilson Observatory on Saturn. The seeing conditions were phenomenal, one of the two best occasions in Livingston's 5,000 hours of observing. Even Lyot's rendition, he wrote, fell short of reality. " . . . the rings were fascinating, appearing as in an imaginative fine engraving. Each major ring was sharply bounded and contained numerous concentric internal subdivisions, resembling dark threads. The 1953 drawing by Bernard Lyot is suggestive of what we saw, except that we would tend to make the subdivisions narrower and of higher contrast." Their view recalled Tuttle's "minutely subdivided" rings of 100 years earlier.

Whom to believe? Lyot and Dollfus? Kuiper? Many could write it all off as more of the confusion and contradiction that had plagued the field for 300 years. After all, they were only visual observations. But Earth-based instrumental observations would eventually confirm the existence of one gap disputed by visual observers. In 1978, Harold Reitsema used a photometer to measure the brightness of the satellite Iapetus as it passed through the shadow of the A ring. During the first intensive observations of such eclipses since 1889 (though they recur about every 15 years), Reitsema detected a slight brightening of the satellite as the Sun broke through a narrow gap or division about one-fifth of the way in from the outer edge of A. Reitsema noted the apparent coincidence of the division with the major feature in Dollfus's rendition of the A ring, and concluded that he had confirmed the existence of Encke's division. He even found a resonance of Mimas that might clear the division. He had indeed verified the reality of the second feature to be generally recognized in the 350-year history of ring observations. Whether it is Encke's division remains controversial. Dollfus, Alexander, and Encke himself had placed that relatively broad feature near the middle of the A ring. Keeler, Lyot, and Dollfus had made a point of distinguishing their narrow black lines near the outer edge from the fuzzier feature toward the middle. Alexander even assigned one of Ainslie's occultation flickers to a gap at the eventual position of Reitsema's gap, while placing Encke's "minima of light" at the center of the ring. The formal designation of the narrow, outer line as the Encke division by an official nomenclature committee will probably stand as a monument to the inevitable controversy surrounding visual observations.

While the controversy of the appearance of the rings from Earth smoldered, most astronomers applied increasingly sensitive electronic instruments to ring problems that now seemed more tractable than those approached through visual studies. By choosing different wavelengths, an instrumental observer could see the rings with different eyes. At the submicron wavelengths of visible light, for example, the rings appeared bright and reflective, but at the micron wavelengths of infrared radiation, they were relatively dark and less reflective. That suggested particles of ice. But which ice? The exotic chemistry of a gas giant like Saturn might have spawned water ice, ammonia ice, methane ice, or a combination of ices. Late in 1969, Kuiper and his colleagues at the University of Arizona thought that the new technique of Fourier transform spectroscopy had given them the answer. To them, ammonia frost absorption of infrared radiation most closely matched the spectrum of the rings that they acquired on 19 and 21 November. Their rings were ammonia snowbanks. Kuiper believed that he had solved the mystery, but others might get the same answer. The cryptic messages of Galileo's and Huygens's day were by then long out of fashion. The press release, however, was the twentieth-century equivalent. Press stories followed on 2 December in the *New York Times* and in the January issue of *Sky and Telescope*.

The Arizona press release made interesting reading for graduate student Carl Pilcher and fellow students in the planetary group at MIT and Hugh Keiffer, a new assistant professor at UCLA. Within a few days this youthful group had decided that Kuiper was wrong and that the best fit was probably with ordinary water ice. By 16 December they had a draft paper saying just that. After the MIT group provided a copy of the manuscript to their competitors at the University of Arizona, laboratory studies in Arizona confirmed the identification. On 5 January Kuiper's group sent a letter to *Sky and Telescope* correcting their identification, explaining that they had not allowed for variations in the spectrum of ice due to the low temperature of the rings. The letter appeared in the February issue, along with an added footnote citing the MIT group's communication. By mid-January the MIT group was presenting their view at the annual Division of Planetary Sciences

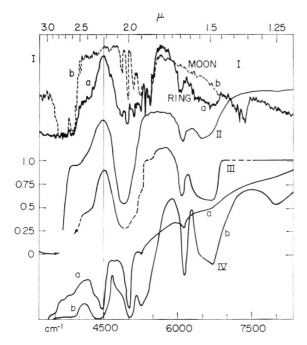

Figure 2.8

The rings are covered with frozen water. After the infrared reflectance spectrum of the rings was recorded, the correct deduction of ring composition required a comparison with the spectra of known substances determined in the laboratory. Curve Ia is the ring spectrum obtained by Kuiper's group. In order to compare spectra obtained by observing through Earth's atmosphere and to remove peculiarities of instrument response, Pilcher and his colleagues divided the ring spectrum by the spectrum of the Earth's moon (Ib) to produce curve III. They then compared curve III with the spectrum of fine-grained water frost—curve II—and spectra of fine-grained ammonia frost—curve IVa—and coarse-grained ammonia frost—curve IVb. The key region for comparison was at 2.2 microns, marked here with a thin vertical line. The peaks of reflectance in the ring and water-ice spectra match well, but the ammonia ice has low reflectance at that wavelength. [Reprinted, by permission, from Carl B. Pilcher, Clark R. Chapman, Larry A. Lebofsky, and Hugh H. Kieffer, "Saturn's Rings: Identification of Water Frost," *Science* 167 (1970), 1372, copyright 1970 by American Association for the Advancement of Science] meeting in San Francisco and submitting it in Washington, DC, to the prominent American journal *Science*, where it appeared in very short order in the 6 March issue. Their identification stuck, at least for the outer rime of frost penetrated by infrared radiation. A slight reddish tint suggested that the ice was dirtied by some contaminant, perhaps rock powder, or had been damaged by radiation.

The rings were water ice. But does the ice come in the size of snowflakes, snowballs, or snowmen? Fifty years earlier Jeffreys had suggested that collisions had ground much of the ring material to a powder. The detection of thermal infrared radiation emanating from the ring particles and the failure to detect emission of centimeter-wavelength radio waves from them tended to support that theoretical conclusion. The infrared emission showed that the particles were warm enough to emit radio waves as well, if 90°C above absolute zero is warm. Since they did not, the alternative seemed to be that the particles were too small—much less than a centimeter across—to emit radio waves efficiently.

The collisions that had pulverized much of the ring material must also have spread the ring particles into a broad layer no thicker than the typical ring particle, according to a theory that Jeffreys worked out in 1947. Particles that collapsed into a ring, he calculated, would continue to lose energy until they piled onto one another, forming a roiling sheet of ice, an avalanche in orbit about Saturn. It would not last long. The ring would spread inward and outward, thinning as it went. It would spread, for perhaps a million years, until particles in adjacent orbits could no longer sideswipe each other. Such collisions happened whenever an inner, faster-moving particle could overtake and hit a particle orbiting just beyond it, the inner particle losing energy and a corresponding amount of angular momentum to the particle in the outer orbit. With less angular momentum, the inner particle would fall closer to Saturn; with more, the outer would drift farther away. In time, the particles could overtake and pass without colliding, and the collision-induced spreading would stop. Such a collisionless ring would be one particle thick, a monolayer 274,000 kilometers across and mere centimeters thick. An astronaut could reach out and put his arm through it, but he would take more than 14 hours at a speed of 60,000 kilometers per hour to orbit once around its circumference.

Rings as thin, powdery snowbanks looked like a good bet until 1973, when Richard Goldstein and George Morris managed to bounce 12-centimeter radio waves off the rings and detect the reflected signal back on Earth using the 64-meter antenna at Goldstone, California.

The rings had shown up on radar, even though centimeter-size particles should not have. Radar had detected particles of at least a few centimeters, if not a few meters, it seemed. How could the rings reflect but not emit radio waves?

As had happened after astronomers saw Saturn through the C ring, apparently paradoxical observations prompted theoretical reconsiderations that painted a new picture of the rings. After further observations over a range of wavelengths, James Pollack, who later teamed up with Jeffrey Cuzzi, worked out what the rings could and could not be like. They still had not eliminated all of the alternatives, but they argued that the observations favored one possibility in particular. The rings they preferred had particles of water ice because ice is a poor emitter of centimeter-wavelength radiation, but can reflect it from its surfaces. Rock would not work. Metal would work too, but it seemed an unlikely material to concentrate from the primordial ingredients of Saturn and its satellites. The ice particles could not be in a monolayer, they said; the rings must be much thicker than the size of a typical particle so that radiation reflected off one particle will have another chance, when it is intercepted by a second particle, to be reflected back toward observers on Earth. Otherwise, the rings would be invisible to radar. The sunlit clouds of Earth are brilliantly white due to the same process of multiple scattering.

Many astronomers studying sunlight reflected from the rings joined Pollack and Cuzzi in their support of a "thick" ring. For one thing, a ring many particles thick seemed the best explanation for the opposition effect, a sharp brightening of the rings in visible light just as Earth catches up to Saturn. When far to either side of the path of the sunlight falling on the rings, they argued, an Earth-bound observer could see both the illuminated surface of particles as well as the shadows that they cast on each other. As Earth approaches opposition, when the Sun is directly behind the astronomer viewing Saturn, the shadows disappear behind the particles casting them, and the perceived brightness of the rings jumps. Other researchers, recalling the inevitability of collision-induced spreading into a monolayer, looked instead to shadowing within the nooks and crannies of the particles themselves to explain the opposition effect. Surface shadowing had its problems, though. Other icy surfaces in the solar system, such as those of the Galilean satellites of Jupiter, did not brighten at opposition nearly as much as the rings. Still, the monolayer/many-particle-thick debate continued in the absence of irrefutable evidence on either side.

If the rings were not a monolayer, photometry at visible wavelengths could also show that the rings are mostly empty space. Judging from the opposition effect, only 1 percent of the rings is taken up by particles, which would make the distance between adjacent particles about four times their own size. Despite this relative emptiness, photometry, along with observations of Iapetus eclipses and stellar occultations, showed that the densest rings present quite an obstacle to anyone trying to peer through them. The faint C ring is no more of an obstacle than the thinnest Earth cloud, but a heavy haze as optically thick as the B ring would block all but the brightest stars in the night sky. Looking through to the other side of the B ring would be something like trying to penetrate 100 meters through a heavy snowstorm.

The rings are not quite like a snowstorm, though, according to Pollack and Cuzzi. They suggested that ring particles could not simply be millimeter, centimeter, or meter size. Instead, the rings must include a range of sizes without the particles being mostly smaller than a few millimeters or mostly larger than a meter. To explain how well the rings reflected the whole range of the microwave wavelengths of radar, they argued that for every 1-meter particle there would be 1,000 10-centimeter particles, 1 million 1-centimeter particles, and proportionate numbers of the sizes between those. Instead of uniformity of particle size, like a pure, sandy beach, a ring's distribution of particle sizes would be like that of a rocky beach having a few boulders, many cobbles, but mostly lots of pebbles. Such an inverse power law distribution, they pointed out, is what theoreticians expect of particles that have been banging into each other for eons, breaking some large particles into many more small ones.

Meter-size boulders did not seem to be the end of the surprises. Photometric observations had detected a peculiar brightening of the A ring off to one side of the midpoint of the front of the ring and to the diagonally opposite side of the back of the ring. The brightenings seemed peculiar because any particles in a bright section of the ring must within a few hours revolve along their orbits into the less bright sections. Yet, the bright quadrants held their positions. The most promising explanation, put forward by Giuseppe Colombo of the Center for Astrophysics and his colleagues, was that the gravity of a few ring particles the size of small hills, ice chunks perhaps 100 meters across, would create wakes of smaller particles. Trailing behind and away from these mammoth ring particles, the denser wakes would brighten the ring when they came to be perpendicular to the line of sight from Earth.

Particles were getting larger and more varied, while the observed thickness of the rings seemed to be getting smaller. The increasing sophistication of observing techniques shrank the 15-kilometer upper limit of the early part of the century to 2 to 3 kilometers by 1970. Some observers believed that they had finally sharpened their methods to the point of actually measuring a true thickness of about 2 kilometers, all done from a distance of more than 1 billion kilometers. Some adherents of the monolayer ring even took a 2-kilometer thickness to mean that ring particles were 1 to 2 kilometers across. There were doubters, however. They viewed the observed thickness as still being only an upper limit of 2 to 3 kilometers; the rings might be much thinner, they said, without approaching Jeffreys's monolayer. Pollack estimated from the photometric properties and the apparent size distribution that the rings might be 100 meters thick. That is hardly thin enough to stick your arm through, but thin nonetheless. That stadium-size sheet of cardboard would have to be no thicker than tissue paper to be as thin.

While observational astronomers were putting together a picture of the rings from the vantage point of a ring particle, theoreticians were grappling with perhaps the grandest problem of all—why are there an A ring and a B ring separated by a gap, and a much fainter C ring inside them all? The answer after 100 years of failing to prove Kirkwood's gap-clearing proposition was much the same—Mimas makes it so. The Mimas 2 : 1 resonance near the Cassini division would, the theory went, perturb any particles orbiting there until their orbits became unstable and they took up more stable ones outside the resonance. By the same reasoning, the Mimas 3 : 1 resonance could shape the inner edge of the B ring, and the Mimas 3 : 2 could form the outer edge of A. Such matching of resonances with ring features seemed rather satisfying, even convincing, when presented in textbooks, but physicists yearned for something more rigorous, more scientific. Their problem was the immensity of the task. Saturn had trapped trillions upon trillions of chips, chunks, and boulders of ice nudged repeatedly if ever so gently by Mimas's gravity as they caromed off each other once or twice a day. Even in the age of computers, calculating the exact physical effects of Mimas and Saturn on the ring particles and their effects on each other was an intractable problem.

The alternative to exact calculation was to assume that the rings were as simple as they could possibly be—particles spread out so far in a single layer that none ever collided. That reduced the calculation to the classic three-body problem—the effect of three bodies gravitationally pulling on each other, one (Saturn) being far more massive than the second (Mimas) and the third (a ring particle) being of negligible mass. Such calculations did not allow any collisions, even when ring

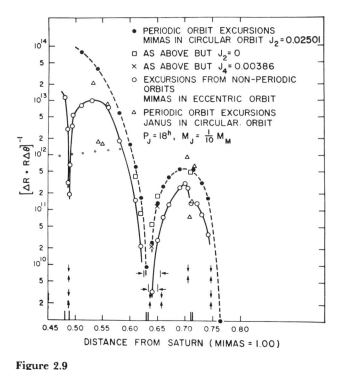

Figure 2.9

How to make rings? Franklin and Colombo predicted the density profile of the rings assuming that they are a monolayer free of collisions. By imposing the resonances of Mimas in a circular orbit (filled circles) and in its true elliptical orbit (open circles), and in addition to those of Mimas the resonances of the proposed Janus (triangles), they created density patterns somewhat resembling the boundaries (marked by the vertical arrows) and densities of the real rings (A, B, and C from right to left). The predicted Cassini division falls somewhat inside its variously reported position indicated by the horizontal arrows and brackets. Although the relative densities of A and B are about right, the predicted C ring is far too dense. [Reprinted, by permission, from F. A Franklin and G. Colombo, "A Dynamical Model for the Radial Structure of Saturn's Rings," *Icarus* 12 (1970), 342, copyright 1970 by Academic Press] particles were driven far from their undisturbed orbits. The set of rings produced under these assumptions by Fred Franklin, Allan Cook, and Colombo in the early 1970s had encouraging similarities to the real rings. The outer edge of their computed A ring fell near the position of the real edge of A, and a gap of sorts opened quite near the position of Cassini's division. Ring B was brighter than A, as it should be. If they stretched the computer model's circular orbit of Mimas into its proper elliptical shape, a gap opened between the B and C rings, as reported by some observers, the Cassini division widened, and some subdivisions appeared in the A ring. They could also add moons besides Mimas. Massive but more distant Titan helped brighten the outer part of B. Tiny Janus, a reported but unconfirmed interloper orbiting between Mimas and the rings, enhanced the brightness variations in the middle of A, that is, in Encke's division.

The computer model of the rings had its problems, however. The faint C ring came out as bright as the B ring. The calculated position of the Cassini division was offset slightly; more worrisome, the width of the calculated Cassini never seemed to measure up to the 4,800-kilometer breadth of the real thing. In fact, in 1977 Richard Greenberg and Franklin decided that neither orbital perturbations nor the resulting collisions could stretch the gap beyond 30 kilometers.

Peter Goldreich and Scott Tremaine suggested another means of widening the gap, one which they had borrowed from the study of galaxies. Spiral galaxies, pinwheellike collections of stars, dust, and gas, had presented a paradox—their spiraling arms are old enough to have rotated perhaps 100 revolutions, but they show only one or two accumulated turns. Unlike a rigid pinwheel, the arms of a spiral galaxy should make more turns the closer they are to the center of the galaxy, which would steadily tighten the spiral. Galaxies do not have tight spirals, so C. C. Lin and Frank Shu concluded that the arms are only waves of more densely packed matter that pass through a galaxy, like a wave passing beneath a boat. Unlike ocean waves, such spiral density waves propagate by their own gravitational attraction for other particles rather than through particle collisions. Goldreich and Tremaine suggested that Mimas could set up the same kind of wave at Mimas's resonance in the Cassini division, despite the immense differences in the size and density of galaxies and rings. Mimas's spiral density wave would carry angular momentum outward, leaving the particles from which the wave took the momentum to drift out of the gap. All in all, the resonance theory of ring structure might need a few adjustments, such as spiral density waves, but satellite resonances seemed to be quite sufficient to explain the rings' appearance.

Not only did resonances explain most of what astronomers saw around Saturn, but classic resonance theory also seemed to explain what astronomers saw, or rather failed to see, around other planets. Researchers heard why resonances might be vital to the very existence of rings at a workshop on Saturn's rings held in 1973 at the Jet Propulsion Laboratory in Pasadena, California. Gordon Pettengill, a radar astronomer and moderator of the post-presentation discussion, asked, "Why is it that only Saturn has rings? Is there a simple one word answer to that?"

"I have an answer," replied Franklin.

"Because it has Mimas?"

"Because it has Mimas, essentially, but I don't think I can say much more. I think the arguments are clear insofar as they are arguments. It is really speculation at this point."

Saturn has Mimas, Franklin argued, and none of the other outer planets—Jupiter, Uranus, or Neptune—have anything like it. Mimas, because of its proximity to Saturn, is the only major satellite that has a strong resonance inside of the Roche limit, the zone within which dynamicists expect to find rings. The limit's namesake, the nineteenth-century French mathematician Edouard Roche, had calculated how close a liquid satellite can approach a planet before the planet's gravity tears it apart. There are no known liquid satellites, of course, but the Roche limit is a convenient outer boundary of the zone in which a satellite might break up to form a ring or material circling a primordial planet might fail to coalesce into a satellite in the first place. Once a ring formed, the resonances of Mimas that reach inside the Roche limit could pen the ring particles within the boundaries seen today. Particles spreading outward into the Mimas 3 : 2 resonance at the outer edge of A, for example, would be thrown into irregular orbits that would allow them to collide with particles still inward of the resonance. Collisions would quickly subdue the orbital irregularities and trap the particles within the ring once again. That would work for Mimas and Saturn, but not Jupiter. The Galilean satellites of Jupiter are larger than Mimas, but lie too far beyond the Roche limit. Franklin and Cook planned to offer this same argument in a paper entitled "On the Absence of a Ring Around Uranus." Pollack, in his 1975 review on the rings of Saturn, concluded from these sorts of considerations that " . . . if the presence of strong satellite resonances within the Roche limit are [sic] a requirement for the presence of rings, it is quite reasonable that Uranus and perhaps Jupiter lack a ring system." Like Saturn's, he might have added.

The uniqueness of the Saturn ring system gained further support from cosmogonic theories, those speculations on the way the present solar system arose from a cloud of dust and gas. According to conventional wisdom, the planets coalesced as this hot, turbulent nebula contracted to form the Sun, and the planets' satellites in turn grew out of the contracting nebulas that formed each planet. An early step in the formation of a satellite is the condensation of gases into small particles. The nature of a satellite, whether rock, ice, or a mixture of both, indeed its very existence, then hinges on the outcome of a race within the nebula. The race is between two competing processes— the agglomeration of the small particles into ever larger ones and the voracious appetite of the growing planet for the newly formed particles. As the gas drags on the smallest particles, they spiral into the protoplanet. If the gases at a given distance from the planet are too hot to condense, or the gases too dense to allow time for agglomeration of particles that do form, there will be no satellite there.

Pollack applied the same reasoning to the special case of ring formation by condensation. Saturn seemed to fit nicely. Condensation of water vapor into ice inside the Roche limit probably continued until there was too little gas to drag all of the ice into the new planet. Jupiter was probably different, Pollack concluded. Radiating ten times more heat as it contracted than Saturn, Jupiter made its eventual Roche zone too hot for the condensation of water. When rock was forming there, the huge bulk of the contracting planet still engulfed the present Roche limit and dragged rocky particles into the planet. The requirement that condensation continue until the planetary nebula nearly dissipates "might not be fulfilled for some of the other Jovian planets," Pollack concluded, "and could provide an explanation for their lack of a ring. This might be the case for Uranus and Neptune. We could claim that Jupiter's nebula never reached the condensation temperature for ice within the Roche limit before dissipation occurred, while the nebula's temperature was always below the condensation temperature for silicates. In this way we could explain the uniqueness of Saturn's rings."

It all fit, or in the favorite phrasing of scientists, all of the observations were "consistent with" theory—there were no loose ends. At the 1973 workshop, Pettengill checked, just to be sure of that. Turning to Franklin, he asked, "Then you feel our understanding is complete in the sense that we think we know why Saturn has rings and also why no other planet has any?"

"I am not completely certain," Franklin replied, "but I am giving possible suggestions that have to be looked into."

After some discussion of the peculiarities of Neptune's satellite system, Pettengill attempted another clarifying question — "There could be rings on some planets that are too thin to be seen optically, I suppose. What are the limits to completeness in the argument?" He never got much of an answer. No one ever mentioned Uranus or Neptune. Franklin and Bradford Smith, a planetary astronomer at New Mexico State University, noted that there had been reports of a ring around Jupiter, usually based on sightings of its shadow on the planet. Smith discounted such reports, having photographed similar features that on closer inspection turned out to be clouds. If a serious, thorough ring search had been conducted and reported by anyone, this group of ring experts seemed unaware of it. Negative results from a search for something that should not be there does not, after all, make for a memorable scientific publication. The tentatively offered, cautiously worded theories of ring formation only reinforced the neglect of ring searches. What experts called among themselves "only arguments" became common wisdom among astronomers. Even the specialists often spoke as if they were only explaining the obvious fact, the established truth, of the uniqueness of Saturn's rings. Carl Sagan was not alone in 1975 when he asked in a *Scientific American* article, "Only Saturn has rings. Why?"

Rings Rediscovered

Since belief in the uniqueness of Saturn's rings was firmly entrenched, the discovery of the Uranian rings came by accident. As with many accidental discoveries, it resulted from the application of a more powerful technique (stellar occultations) to a new subject (Uranus). Although astronomers had been observing occultations of stars by planets sporadically for centuries, they had never seen a Uranian occultation before. Furthermore, it was only during the preceding 25 years that they had the photoelectric instrumentation readily available to learn the temperatures of planetary atmospheres from stellar occultations.

The German astronomer Pannekoek launched the modern history of occultations when he visually observed the occultation of a star by Jupiter in 1903. He realized that it is not obstruction by the planet's atmosphere that dims a star, but rather increasing atmospheric refraction—the bending of starlight as it passes through denser regions deeper in the atmosphere. The same effect, on a smaller scale, causes the Sun to appear "squashed" when observed near the horizon. By accurately measuring the starlight intensity during an occultation and knowing what gases the planet's atmosphere contains, one can calculate the temperature of the planetary atmosphere.

This would have been a wonderful technique for obtaining the atmospheric temperatures of distant planets, but two factors prevented its application until several decades later. First, an observer must measure the starlight to an accuracy of a few percent at high time resolution—about once every 0.1 second is the slowest that can be ordinarily tolerated. This became technically possible with the invention of the photomultiplier, a light detector that combines high precision with fast response. The second problem was knowing when occultations would occur—preferably the prediction should come well in advance

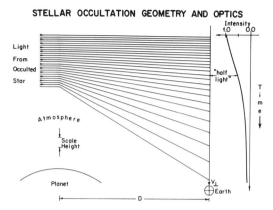

STELLAR OCCULTATION GEOMETRY AND OPTICS

Figure 3.1
Starlight dims by refraction. In this exaggerated diagram, light from a distant
star is represented by parallel rays, incident from the left. Since the density of
the atmosphere increases rapidly (exponentially) with depth, each successive ray
of starlight passing through it is bent (refracted) more than the one above. This
causes the observed intensity of the starlight to decrease as the Earth passes
through the pattern of refracted starlight, as shown on the right-hand side of the
diagram. For the case shown here of an atmosphere whose temperature and
composition do not change with altitude, one can calculate the temperature of
the atmosphere from the rate that the starlight dims. [Reprinted, by permission,
from James L. Elliot, "Stellar Occultation Studies of the Solar System," *Annual
Review of Astronomy and Astrophysics* 17 (1979), 445, copyright 1979 by Annual Reviews Inc.]

in order to give time for the necessary preparations. Before the age
of computers and reliable star catalogs, no one systematically predicted
occultations — at least they did not publicize them well.

Bill Baum and Art Code took up Pannekoek's suggestion when they
observed the occultation of the star σ (sigma) Arietis by Jupiter in
1952. However, their analysis pointed to an impossibly high temperature. It was the first attempt to measure the atmospheric temperature
of a planet with an occultation, a great step forward, but the implausible
result, coupled with the scarcity of events, dampened any enthusiasm
for using occultations to learn about the planets.

The next predicted occultation was that of the bright star Regulus
by Venus in 1957. Although the occultation was widely observed
visually from Europe and Africa, observers at only two stations obtained
photoelectric data. Unfortunately, they could not reconcile the 310°K
temperatures (0°C = 0° Celsius = 273° Kelvin = 273°K) obtained
from the analysis with those inferred from atmospheric models for

Venus based on spacecraft data. Again the promise of atmospheric temperatures from stellar occultations faded. However, Gordon Taylor of the Royal Greenwich Observatory and D. Ya. Martynov of the Sternberg Astronomical Institute independently put the occultation to another use by calculating an accurate diameter for Venus, using a selection of the more than 200 visual timings of the disappearance and reappearance of Regulus—an important step forward in spite of the difficulties with the atmospheric temperature.

In 1968 it was Neptune's turn to occult a star, BD—17°4388, and observers successfully recorded the event in Australia, New Zealand, and Japan. The data yielded the diameter and oblateness of Neptune, as well as the temperature of its atmosphere. Independent analyses of the different data sets gave different temperatures. While some results came close to what should be the correct value, 140°K, others were seriously discrepant. Occultations might work as advertised, but high-quality data and careful analysis were needed to get sensible results. An exciting new feature of this event, noticed because this was the first occultation for which the data were recorded at moderately high time resolution (0.1 second), was the presence of numerous sharp "spikes" or bright flashes of light that occurred while Neptune occulted the star. Some discussion then ensued about whether the spikes were actually produced by Neptune's atmosphere or were erroneous artifacts introduced into the data.

Only 3 years after the Neptune event, in May 1971, Jupiter would occult the second-magnitude star β (beta) Scorpii, according to a prediction by Gordon Taylor. At Cornell University, Joe Veverka wanted to observe the occultation from Boyden Observatory at Bloemfontein, South Africa. He planned to observe simultaneously at three different wavelengths and to record the observations at high time resolution (0.01 second), so that the spikes—if real—would show up clearly. Not being an instrumental type, Joe enlisted his former Harvard colleagues, Bill Liller and me, to build a three-channel photometer and the electronics needed for the high-speed data recording. The project seemed interesting, so I accepted his invitation, even though I was trying to finish my Ph.D. research at the time. The fourth member of our team, Larry Wasserman of Cornell, joined us with the idea of developing a thesis topic from the data that we hoped to obtain. At the University of Texas, Bill Hubbard and the lunar occultation pioneer David Evans planned to dispatch observing teams to sites in South Africa, India, and Australia. A French team, from the Observatory at Meudon, also planned to observe from South Africa.

Most of the observations of this occultation succeeded. Shortly thereafter, I finished my degree and moved to Cornell to help Joe and Larry analyze our data and to build better photometric equipment for observing occultations. All groups working on the β Scorpii occultation agreed that the temperature of the Jovian upper atmosphere is 170°K. In addition, our multicolor data at high time resolution proved valuable for another purpose: to obtain the proportion of helium in the Jovian atmosphere, using a method proposed by Robert Brinkmann of the Kitt Peak National Observatory. Our analysis of the atmospheric composition gave a helium fraction of 0.16, a value with a large uncertainty $(+0.19, -0.16)$, but consistent with more accurate results obtained later from the Voyager spacecraft. With better occultation data, the method could prove a valuable tool for measuring the composition of planetary atmospheres. Bill Hubbard and Tom Van Flandern also determined the diameter and the oblateness of Jupiter from this occultation. We were even able to obtain the angular diameters and separation of the two stars that compose the binary star β Scorpii. These measurements came from a careful analysis of the spike widths: waves in Jupiter's atmosphere were being used as a giant telescope to focus the starlight, analogous to the dancing bright ridges of sunlight, focused by surface waves, on the bottom of a swimming pool. We also showed that we could use our multicolor observations to correct for the light from Jupiter's edge, which had the advantage over other methods of not sacrificing the high time resolution. The results of the β Scorpii occultation made most of us more confident that we could obtain reliable temperatures from occultations. It also revealed that the spikes could provide even more interesting information about the planetary atmosphere and the occulted stars.

Meanwhile, I had finished our new photometer and recording equipment. We used it to observe a series of mutual occultations and eclipses of the Galilean satellites of Jupiter during the fall of 1973 and the lunar occultations of Saturn's five largest satellites—Titan, Rhea, Iapetus, Dione, and Tethys—from Hawaii's Mauna Kea Observatory during March of 1974. Based on this success, we then planned to observe two lunar occultations of Neptune in 1975 in order to learn about the particles in its upper atmosphere. The first occultation, visible from Australia, would occur in August, and the next occultation, visible from South Africa's Boyden Observatory (again), would occur in September. We succeeded in obtaining support for this project from NASA and the NSF (National Science Foundation). Our observing team for the occultations consisted of Joe Veverka, myself, and Ted Dunham, who had entered graduate school at Cornell in the fall of 1974. Joe learned of Ted's interest in instrumentation and had consequently signed him up to work in our lab. Bill Liller would work with us in Australia, but would not be able to continue to South Africa. The Australian observations would be in collaboration with Ken Freeman, who, although not a planetary astronomer, had developed an interest in occultations from his Neptune work in 1968.

Compared with our efforts for the β Scorpii occultation, the Neptune occultation required a lot more work. We had more elaborate (and more massive) observing equipment that we would set up at two observatories—on three telescopes at Mount Stromlo and two at Boyden. However, the odds of success were heavily stacked in our favor. We estimated about a 50 percent chance of reasonably clear weather at Mount Stromlo and a 65–70 percent chance of clear weather at Boyden. Contrary to other types of astronomical observations, lunar occultations require clear skies for only a few seconds around the time of the event. Hence, we had better than an 80 percent chance of getting the instant of clear weather that we needed for at least one of these occultations.

The evening of 15 August began clear at Mount Stromlo, but thick clouds soon began to pour in. They blocked observations in Australia. Disappointed, we packed up our gear and moved on to South Africa, where we could see the moon occult Neptune on its next monthly pass through that region of the celestial sphere. On the night of the occultation, we became a bit uneasy because days of clear skies had given way to "occultation weather," the term that we used to describe a sky partially covered with moving cumulus clouds. The sky was clear in the holes between the clouds, but unluckily the occultation occurred behind a cloud. Months of work had ended up in a total loss. I remember the glum conversation that Ted and I had in the observatory kitchen after we had closed up the telescopes. We talked of new projects we might try—these occultations were fine when the weather was clear, but just too risky to invest so much effort and end up with nothing.

After returning home to Cornell in September, my thoughts turned to a quasar project that I had in mind. I had requested telescope time for it and asked the NSF for support. Our bad luck had also discouraged Joe, but he had been thinking ahead to the occultation of ϵ (epsilon) Geminorum by Mars, which would occur on 8 April 1976 and be visible from the eastern United States. West of the Mississippi River the Sun would not have set by occultation time. My thought was that east coast weather in the spring was liable to be cloudy; we had had enough misery. It was not worth all the effort of preparing for the observations just to be clouded out again. Joe's point was that we should certainly try *something*, since we had been supported for occultation work and the Australia-South Africa debacle would not have impressed our sponsors. Neither of us could convince the other. An exciting idea unexpectedly resolved the standoff: we would propose to use the Kuiper Airborne Observatory for the ε Geminorum occultation. We could observe without being clouded out, which satisfied my objection. Fortunately, all my efforts could be devoted to preparing for the airborne observation because NSF did not approve my proposal for the quasar project—although I could not take it so philosophically at the time.

Since requests to use the KAO greatly exceed the available flight time, a NASA review committee rates the proposed research programs, ranks them, and grants observing flights to those at the top of the list. In defining the objectives of our proposed research on the KAO, our thinking was influenced by the impending Mars landings of the Viking spacecraft, the first being scheduled for 4 July 1976. We could measure the temperature of the upper atmosphere and compare these results directly with data from the Viking entry probe. If successful, we could establish once and for all the reliability of the occultation method of measuring temperatures. Potentially more important for the Viking mission, however, would be the opportunity to determine the composition of the Martian atmosphere. Although it had long been known from spectroscopic observations that carbon dioxide was a major constituent, the amount of argon was unknown. Argon posed a potential problem for the gas chromatograph aboard the Viking spacecraft, since a large amount would destroy it.

In our request, we needed to establish that our scientific objectives would require the unique capabilities of the KAO and that ground-based observations would not do just as well. We emphasized that the occultation was a one-time-only event and that we would run a great risk of being clouded out—particularly on the east coast of the United States in April. Furthermore, being above 82 percent of the Earth's atmosphere at 41,000 feet would eliminate virtually all of the noise in our data from atmospheric turbulence. This reduced noise would translate into more accurate measurements for the temperature and composition of the Martian atmosphere. Hence, we should expect substantially better temperature profiles and other results from the KAO than could be achieved from the ground—even with a larger telescope.

To us, our case seemed strong, and we were optimistic about the success of our proposal. Nancy Boggess, the chairman of the review committee, gave us a rude shock when she called me to report the results of the review. She had submitted the proposal to 14 non-NASA scientists: 2 were in favor and 12 were against. Therefore, NASA could not support our observations with the KAO. The reviewers against us ranged from those who thought that we could not obtain good data from the airplane to those who thought that even if we did, our results would be irrelevant. Joe and I were absolutely convinced that we could obtain excellent data—better than any previous occultation—and that our results were of potential importance. We were not about to take no for an answer!

We enlisted the help of our boss, Carl Sagan, who appreciated what we were trying to do and had the ear of more influential people at NASA than we did. A major factor working against us was that we were trying to break into the KAO schedule in the middle of the year, so that our project was being judged by particularly stringent standards. Furthermore, the planetary community did not yet fully appreciate the powerful potential of occultations for probing the planets. The wider astronomical community, from which most of the referees of our proposal were drawn, was even less aware. In the end, the arguments in favor of our work won—probably because of our potential impact on Viking. Just 3 weeks before we would be scheduled to fly, Nancy Boggess called me with the good news. Now the ball was in our court. With all the uproar that had surrounded our proposal, we had to succeed—to vindicate ourselves and those who backed us.

In order to achieve the best results for the argon measurement, I decided that we should use the mobility of the KAO to observe from a point on Earth where the star would follow an apparent track *perpendicular to the edge of Mars*, both when it went behind the planet and when it reappeared out the other side. We could achieve this configuration if we chose our course to put us directly in line with the center of Mars and the star at the time of midoccultation, that is, in the very center of Mars's shadow.

Since the ε Geminorum occulation would be visible in the eastern United States, our flight plan called for a run across the country on the afternoon of 8 April from the KAO's home base at NASA's Ames Research Center south of San Francisco. We reached the east coast at dusk and then changed to a course that would allow us to track Mars. It soon appeared on the television screen that monitors our photometer, with ε Geminorum right alongside. Since we were tracking Mars, it stayed in the center of the field, and the star followed a slow course toward the planet. Soon the images merged, although Mars had not yet occulted the star. Attention was riveted on the chart

Figure 3.2
Mars and companion on television. As Mars (left) moves steadily to occult the
star ε Geminorum, a television monitor displays images of the two bodies from
light picked off by a beam splitter in our photometer.

recorder while we waited for the signal to drop, which would signify
that the occultation was in progress. Right on schedule, the signal
dropped, but only a few spikes appeared, not nearly so many as we
had observed in the occultation of β Scorpii by Jupiter. The star was
due to reappear five minutes later, but after only 2 1/2 minutes, the
signal rose and dropped sharply within a few seconds. Had something
gone wrong? Did Mars have a hole in the middle?

No. I quickly realized that by passing directly behind the center of
Mars, we must have observed the focusing of starlight around the
planet by the entire atmosphere of Mars! This newly observed phenomenon certainly surprised us, and we decided to call it the "central
flash." To observe the central flash, as we subsequently calculated,
required that we be within 25 kilometers of the exact alignment of
the star and the center of Mars. This was fine testimony to the accuracy
of Gordon Taylor's prediction and Bob Morrison's navigation. Now
we could relax, knowing that we had the data that we wanted so
much and had the bonus of the central flash to boot.

At Cornell, we began data analysis immediately, since we wanted
to give all the information that we could to the Viking project. By late

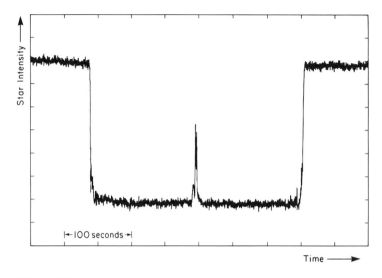

Figure 3.3
Does Mars have a hole in the middle? This graph shows the combined light intensity of Mars and the star, plotted versus time on the horizontal axis, that we measured from the Kuiper Airborne Observatory. The initial drop in intensity occurred when Mars occulted the star, leaving the baseline level of light from Mars alone. Midway through the event, the signal unexpectedly rose for a few seconds and then dropped back to the signal from Mars. This brightening—the "central flash"—was caused by the focusing of starlight by the Martian atmosphere, all around the edge of the planet. The central flash had never been observed before because it is visible from only a small area on Earth near the path traced by the line passing through the star and the center of Mars.

June, Dick French, Peter Gierasch, and I had concluded in a report to the Viking project that the average temperature of the Martian upper atmosphere was 140°K and that the temperature profile contains large wavelike structures, which we later decided were caused by tides due to daily heating of the surface of the planet. Richard Zurek and Conway Leovy of the University of Washington had predicted such tides on Mars. Our argon measurement did not turn out to be as precise as we had hoped because the light curve did not contain the large number of spikes that had appeared during the occultations by Jupiter and Neptune. We could set an upper limit of 30 percent on the amount of argon. Fortunately, the amount of argon in the Martian atmosphere turned out to be only about 2 percent and did no harm to the Viking experiments. The central flash proved to be not only exciting to observe, but we also figured how to use it to measure the average global extinction of starlight by the atmosphere in a region just below that for which we obtained temperature profiles. Observations at a number of ground-based telescopes, although of lower quality than the KAO data, were combined to calculate the shape of the Martian upper atmosphere and strengthen the case for atmospheric tides as the cause of the wavelike temperature structure. The agreement of our results with subsequent Viking results proved that stellar occultations can provide a variety of reliable information about the upper atmospheres of planets and that the KAO is a powerful observatory for occultations.

It was not long before these conclusions were put to the test. In 1973, Gordon Taylor had predicted that Uranus would occult a relatively bright star—designated SAO 158687—on 10 March 1977 and that the event would be visible from areas surrounding the Indian Ocean. Taylor had identified this first predicted Uranian occultation from his systematic search of the Smithsonian Astrophysical Observatory (SAO) catalog for occultations by planets. Several groups saw this event as a good opportunity to measure the temperature of the Uranian upper atmosphere and the diameter and oblateness of the planet. Taylor predicted that the occultation would be nearly central; that is, the center of the shadow of Uranus would pass over the Earth. Since Uranus is over four times the size of the Earth, the occultation would be visible wherever it would be dark and Uranus would be above the horizon.

At Cornell, we had this event in mind for our next good occultation observation, particularly since it was the first one for Uranus. In March 1976, during our heated campaign to obtain use of the KAO for the ε Geminorum occultation, we had requested the KAO for the Uranian event. As stated in our proposal to NASA, "This occultation offers a unique opportunity to learn about the atmosphere of Uranus, as well as the size and oblateness of this planet. . . . Since Uranus is so large, there's virtually no chance of missing the occultation. The best strategy for airborne observations would be to take off from Australia, to observe the occultation over the Indian Ocean and then return to Australia."

This time the approval of our application went smoothly. First of all, there were no longer doubts about the value of occultations or the advantages of the KAO for observing them. Second, and equally important, many infrared observers wanted to use the KAO in the Southern Hemisphere to observe objects inaccessible from northern skies. With the Uranus occultation as the focal point, the airborne review committee recommended that the KAO embark on a scientific mission to the Southern Hemisphere in the spring of 1977.

As preparations got underway, we began working closely with the group at Lowell Observatory. This tie developed because Larry Wasserman, after graduation from Cornell, took a position at Lowell, where he continued occultation work with Bob Millis, who had developed an interest in the field from the Galilean satellite occultations and eclipses in 1973. Bob obtained spectra of Uranus and SAO 158687 in the summer of 1976. His spectra showed the methane absorption bands in the spectrum of Uranus becoming progressively deeper toward the longer wavelengths. Hence, the best wavelengths for the observations would be at far red wavelengths, where the intensity of the star relative to Uranus was the greatest.

Our plan was straightforward. Bob would observe the occultation with the 0.6-meter planetary patrol telescope at Perth, and we would observe with the KAO on a short flight west of Perth. In addition, Bill Hubbard and his colleagues at the University of Arizona planned to field teams of observers at several sites, including Australia and India.

For major occultation efforts, I had developed a pattern of carefully thinking through all aspects of our plans in order to identify where we might go wrong. This had become particularly important for the Uranus occultation, since it involved such a large expedition. By December I began thinking that it would be a good idea to check the relative position of the star and Uranus, since the observing time on the KAO would be limited; I did not want to arrive at the predicted location and find that the occultation had started without us. In his original prediction of the occultation, Gordon Taylor had pointed out that Uranus would pass close to the star during January before reversing its motion and occulting the star in March. I talked to the people at Lowell about the possibility of taking this opportunity in January to refine the prediction, since they had the telescopes and measuring equipment to do the job. They agreed that a refinement would be prudent; Larry Wasserman and Otto Franz set to work on it.

In late January, Larry called with some surprising and upsetting news. The position of the star that he and Otto measured differed considerably from the catalog position that Gordon Taylor had used to calculate the predicted path of the shadow of Uranus. Their new position for the star indicated that the shadow would only graze the southern part of the Earth, or even miss it completely. Hence the occultation would only be visible from extreme southern latitudes— if at all. Could there be an error in the new star position? Larry and Otto continued to measure photographic plates taken both at Lowell and at the US Naval Observatory in Flagstaff. The calculation was carefully checked. The average of their measurements became more

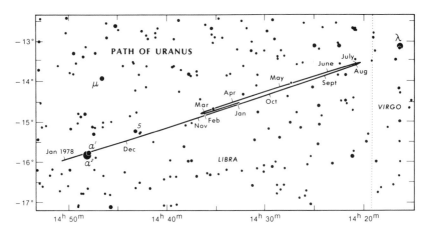

Figure 3.4
The path of Uranus through the stars. In January 1977, Uranus passed nearly in line with the star designated SAO 158687 before reversing its motion and returning to occult it in March. The apparent looping motion of the planet as seen from Earth, known as a "retrograde loop," is due to parallax and the faster velocity of the Earth in its orbit. The January conjunction gave Larry Wasserman and Otto Franz at Lowell Observatory an opportunity to measure carefully the relative position of the two bodies and to calculate more precisely where the March occultation would be visible. They concluded that the occultation would occur only at high southern latitudes, with the possibility that the shadow of Uranus would miss the Earth altogether. [Reprinted, by permission, from *Sky and Telescope* 53 (1977), 51, copyright by Sky Publishing Corporation]

accurate, but the conclusion remained the same. Meanwhile, other measurements of the star position by observers at Perth and Sydney put the shadow even further south, which implied a complete miss.

Now I had to make a decision—whether to continue with our plans for the KAO expedition or to withdraw. To withdraw would virtually eliminate the possibility of obtaining excellent airborne occultation data for Uranus in the foreseeable future. This was the only occultation predicted for Uranus for the next few years, and apparently no SAO stars had been occulted during the past 25 years. On the other hand, it would be irresponsible to continue with an expedition that had little chance for success.

To resolve this dilemma, I had to estimate the odds for successful occultation observations and then weigh the rewards of success against the costs of failure. Having worked closely with Larry for several years at Cornell and learned of Otto's skills in star position determinations, I was inclined to take their result at face value. They reinforced this opinion during telephone conversations, in which they explained every detail of what they had done to reach their conclusion. I then sought the opinions of Brian Marsden, of the Smithsonian Astrophysical Observatory, and Ken Seidelmann, of the US Naval Observatory, both of whom agreed that the Lowell result and its estimated error should be the best information available. The Lowell prediction called for the northern edge of the shadow to pass north of Perth, with an expected uncertainty of 2,500 kilometers north or south. Hence, Bob had better than a 50 percent chance of observing the occultation from Perth. If we could observe the occultation from the KAO at 50° south latitude, the odds would be 5 to 1 in favor of being within the shadow. With such a good chance for success, I decided to continue our expedition as planned with one important exception: the KAO should fly as far south as possible. In retrospect, I would not have submitted a new proposal that had a 17 percent chance of total failure, and I doubt that the review committee would have accepted that risk either. But since preparations had gone so far, my criteria (and NASA's) had changed.

With that decision behind us, Ted, Doug, and I began testing and packing our gear for shipping to the Ames Research Center. Upon arrival we found major preparations underway for the Australian expedition under the supervision of the expedition leader, Carl Gillespie, the facilities manager for the KAO. Since we would be the first of the four different instrument groups to observe from the Southern Hemisphere, our instrumentation was installed at Ames and we traveled with the KAO. To provide for proper flight crew rest, the trip was divided into a series of single hops, the first to Honolulu, the second to Pago Pago, Samoa. The next day it was on to Melbourne, where the infrared gear was unloaded, since this was where the KAO would return for the other research flights after we completed our work in Perth. In Melbourne, we picked up Wilson Hunter, the NASA representative in Australia, who had made many of the arrangements for the expedition. The flight to Perth was relatively short compared with the long hops that had become a daily routine with us. The entire entourage of the KAO was based in the Sheraton Hotel in downtown Perth.

In Perth, the public interest in the KAO and why it had come all the way to Perth was high. The plane was open for tours and to the press; members of the press wanted to come along on our occultation flight, but in the interest of keeping down confusion, we politely refused. We contacted Bob Millis, who showed us his facilities at the observatory in Bickley, a suburb of Perth. Bob had his photometer on the 0.6-meter telescope, and Ben Zellner, from the University of Arizona, was set up on the 0.4-meter telescope. The five of us got together for lunch at the Pizza Hut, where we again rehashed the prospects for the shadow of Uranus hitting or missing the Earth. No new information had been added to the prediction—it was just "misery loves company." Although the atmosphere was friendly, there was an undertone of "us and them," since we were closely cooperating with Bob and his colleagues at Lowell but were in some competition with the Arizona and other groups. After that meeting we became totally engrossed in our own observing preparations. However, we made arrangements to meet with Bob after the event to compare our results. The quickest way to accomplish this would be for Bob to meet us upon landing at the airport.

On 8 March, we had a calibration flight, during which we measured the respective signals from the star and Uranus while they were still far enough apart to do so, and generally checked out our systems. All seemed in order. Next our attention turned to the plan for the occultation flight itself. In order to have the greatest chance of being within the shadow of Uranus, we should go far south, to the "bottom" of the Earth. However, the maximum flying time of the KAO is 12 1/2 hours. Safe operation of the airplane dictated that a fuel reserve be maintained in order to reach an alternate landing site should Perth be fogged in. In that part of the world, alternate airports for C-141s are scarce, so we had special permission for a single use of Meekathaera, which was rated below the weight of a C-141. The criterion set by the flight engineer was that our reserve should be sufficient to reach our alternate if we should lose one engine at our farthest point from home.

This set our flight time at 10 hours, but still left room for fine tuning our strategy. Since the telescope views the sky perpendicular to the flight path, the course flown by the airplane determines the direction around the horizon that the telescope is pointed. Uranus would be nearly due south during the occultation, so that the tracking leg of our course would run approximately west to east. This meant we could trade off distance south for time that we could record data; the farther south that we went, the shorter the tracking leg on Uranus and the star. The large error in declination ("celestial latitude") of the star dictated that we go south, but an error in its right ascension ("celestial longitude") would mean that the occultation would occur early or late, depending on the sense of the error. In the end I decided, and my team concurred, that we should start observing Uranus 1 hour (a nice round number) before the predicted time of midoccultation (21:06 UT). This implied that we should turn onto the course for

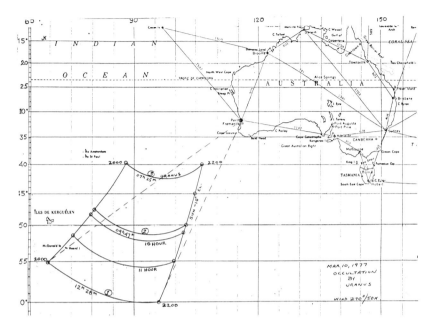

Figure 3.5
How far south should we go? Navigator Jack Kroupa drew this planning chart to
show the trade-off between the distance south we could go and the length of the
flight. I chose the "10-hour" option as being the farthest south that would be
consistent with the safe limit on air time imposed by the flight engineer. The
bottom section of the flight paths show the routes that would be flown to observe Uranus. These are curved because the plane must fly in the direction that
puts the planet abeam in order for the telescope to keep it in view.

tracking on Uranus at 20:00 UT, about 47 minutes before the time
that the star was predicted to disappear behind the planet. Navigator
Jack Kroupa planned a course that would take us to 51°50' (51 degrees
and 50 minutes) south latitude, the farthest south that the KAO has
been before or since.

As the time for our occultation flight approached, my mind drifted
toward thoughts of impending doom. With all the work that everyone
had put into these observations, I was hoping hard that it would not
be a bust. The annoying thought was that success or failure was already
predetermined, but our positional measurements for Uranus and the
star just were not good enough to tell us the answer. In any case, the
suspense would be over in a few hours. As we were boarding the
plane after the preflight briefing, I encountered Don Oishi, one of the
telescope operators who was not scheduled to fly but was coming
along anyway. I asked him why he had decided to come on the long,

overnight flight instead of something more relaxing. "I have the feeling that something important is going to happen tonight," he replied, "and I want to be part of it." Inside, I cringed with thoughts of the shadow of Uranus passing by us, far off the Earth.

I need not have worried. Just a few minutes late, the signal began dropping, as Uranus's atmosphere snuffed out the starlight. And Oishi's premonition was correct: those extra dips proved to be "something important" all right. While the star was behind Uranus, thoughts turned to notifying other observers of the probable "satellite belt." For the preoccultation "satellites" that we had already detected, it would be too late, but others might catch the postoccultation events — if they would leave their equipment turned on for about an hour longer. In Australia, dawn would have occurred, but in South Africa, they would have a chance. It was particularly important to let them know, since they could not have observed the events prior to the occultation by Uranus and hence should have no reason to keep recording data longer than a few minutes after the star reappeared from behind Uranus.

How could I get the information to the South African observatories? There was not much time, and I did not have the telephone numbers of the observatories with me. The name that came to mind was Brian Marsden, the editor of the Circulars of the International Astronomical Union (IAU), who could certainly pass a credible message to the observatories quicker than anyone else that I knew. I drafted a short note that read, "Please inform South African observers immediately that secondary occultations of SAO 158687, presumably by small bodies in orbit around Uranus, were observed from the Kuiper Airborne Observatory. Suggest that they observe until dawn. Jim Elliot."

Carl Gillespie took the note to copilot Dave Barth, who contacted Bob Barrow at Perth by radio. Fortunately Bob was able to reach Brian on the telephone. To relay the news, Brian called J. Hers in Johannesburg, who informed him that it was raining. He then issued an IAU circular, dated 11 March 1977, that read, "R. Barrow, Gerard P. Kuiper Airborne Observatory, has relayed word from Perth of successful observations by J. L. Elliot in the southern Indian Ocean of last night's occultation of SAO 158687 by Uranus. A secondary occultation [sic] was also observed, this presumably being caused by a small body (not [the satellite] Miranda) in orbit about Uranus. J. Hers reports that heavy rain prevented observations in the vicinity of Johannesburg." It was not my intention that Brian publish my message in a circular, since at the time — still in the middle of making the observations — I was not really ready to commit us in print to a satellite belt around Uranus.

After the star reappeared from behind Uranus and more brief occultations had been observed, the sky began to brighten in the east; at 22:17:40 UT we terminated our data recording, since it had become too bright to observe. We then headed north to Perth. Although many on board used this opportunity for some well-deserved rest, I was too excited. Below us I could see a fairly thick cloud cover that boded ill for Bob's observations at Perth. When we landed at 8:43 A.M. local time, Bob was waiting for us at the fence that separated the flight ramp from the public area. I greeted him with "How many satellites did you see?" "At least one, maybe more," he replied. However, there had been no occultation by Uranus at Perth. The "top" of the shadow had passed somewhere between Perth and the KAO.

Bob recounted his experiences at Perth, where he was assisted by Peter Birch and Dan Trout of the observatory staff. He too had begun his observations about an hour early for fear of missing the occultation due to an erroneous prediction. About 10 1/2 minutes after he had begun recording, a 30 percent dip occurred in the signal. The drop looked quite unlike what he had expected for the planetary occultation or for the passage of a cloud, which would have been visible in the moonlit sky. At first, it seemed as if Uranus had drifted partly out of the aperture, but as Bob was preparing to check, the signal suddenly jumped back to the normal level. A quick look in the photometer revealed that Uranus was precisely in the center of the field. The dip in the signal had lasted just over 8 seconds. During the next 14 minutes, he saw four more drops, all of which were shallower than the first and lasted only about 1 second each.

We all went to my hotel room and unrolled our chart recordings to compare our results. Sure enough, we both had recorded brief secondary occultations. The unwieldy lengths—nearly 100 feet—of our chart records, as well as their different scales in both time and signal level, hampered the comparison somewhat. Lack of sleep and excitement also contributed to a less thorough comparison than we might have made at this time. We discussed possible causes of these events—small satellites or rings—and dismissed the ring hypothesis because the events lasted such a short time. Our main objection to narrow rings was that none of us knew any examples of narrow rings. How could Uranus have rings only a few kilometers wide?

Although the satellite belt idea seemed the most plausible, one piece of evidence did not fit. The occultations were not total. If an individual moon caused each dip, one had to assume that we observed grazing occultations in which the starlight would have been only partially blocked by the satellite. This explanation for the absence of total occultations, although highly improbable, was not impossible. In any case, it seemed more likely than narrow rings. If our chart records had the same scale, we probably would have noticed the similar shapes and time intervals between the dips observed from the KAO and Perth.

Wilson Hunter and Carl Gillespie realized the significance of our discovery as well as the importance of publicizing the result. They arranged for a press conference at our hotel in the early afternoon and a colloquium at the University of Western Australia shortly thereafter. Bob had gone back to his hotel, so I called him to say that if he were awake enough and interested, he should come to these events so that we could make a joint announcement of the discovery. At both we explained what we had observed and our interpretation of our results as a "satellite belt" or "swarm of satellites" around Uranus. The story carried the next day on page 5 by *The West Australian* opened, "Two American astronomers working from Perth yesterday discovered a swarm of satellites around the planet Uranus—a find which they say is of major importance to astronomy." It continued, "Dr. J. Elliot, of Cornell University in New York, said that the swarm could contain up to 100 satellites. . . . Dr. Millis identified a satellite 100km in diameter and Dr. Elliot identified several other smaller ones with diameters between 30km and 40km. . . . Red-eyed from lack of sleep, they told a Press conference that the [100km satellite in the] swarm would have been visible only within a range of 50km north and south of Perth or perhaps from a South African Observatory." At the physics colloquium, we displayed the entire KAO chart record in front of the audience. Bob's colleagues in attendance from the Perth Observatory were pleased to learn of our surprising result, since they had assumed that the exercise had been a failure because no occultation by Uranus had been observed at Perth.

By now we felt fairly certain of our interpretation of the data and wanted to gain whatever credit we deserved by reporting our findings to a recognized scientific publication. We did not know who else had been successful in observing the occultation, what their data contained, or what conclusions they had reached. Because we had the KAO data, which showed the secondary occultations from both sides of Uranus, as well as the Perth data, the chances were that no others were as certain of their interpretation as we were. But our advantage would diminish as time went on. We wanted to be "fastest with the mostest"— but not to be *wrong*. The best course of action seemed to be to telephone a summary of our data and conclusions to Brian Marsden for publication in the IAU circulars. We composed a joint communication that afternoon, and I called Brian with the news.

The next day Ted, Doug, and I left Perth while Bob stayed on for further photometric observations of Uranus that might determine its rotation period. We went as far as Melbourne with the KAO, which then began its series of infrared missions. Flying straight through and crossing the international date line, we arrived in Ithaca on the evening of 13 March. All the while, I kept thinking about what our data really meant. The satellite belt interpretation kept bothering me, since all our events would have to have been grazes, or maybe the dips were not total because of some diffraction effect that I did not understand. The following evening, after a day of rest, I invited Joe Veverka and Jay Goguen over to show them the data and to tell them what had happened. We discussed the satellite belt interpretation and the possibility of rings—but narrow rings seemed farfetched, again because narrow rings had not been observed before.

After they left I thought more about the ring idea. If we had really seen narrow rings, then symmetry demanded that the spacing between the dips that occurred before Uranus occulted the star should agree with those after Uranus occulted the star. To dispose of the narrow-ring hypothesis once and for all, I decided to compare the spacings of the dips on the chart. With the help of my wife Elaine, I unrolled the chart paper in our living room and folded it back over itself in order to compare the spacing between the dips. Nearly a perfect match! Uranus is encircled by five narrow rings!

The next day I showed the result to Ted and Doug. What I had done with the chart was only geometrically approximate, since we were not looking at Uranus exactly pole on. From our vantage point the rings would look like a bull's-eye pattern seen from an oblique angle. To confirm the idea of rings, we had to do the geometric calculation correctly to account for our angle of view. Ted converted our time scale to radius in the ring plane and Doug plotted the data on the correct scale. The agreement was even better. It was time for another call to Brian.

Although the conclusion of narrow rings seemed inescapable to me, Brian was skeptical and would not publish it in the IAU circulars—at least not right away. This annoyed me, since just a few days earlier he had published a secondhand telephone message announcing the discovery of a satellite that I had not intended him to publish; now he was refusing to publish a direct report of narrow rings based on solid evidence. On his own, he proceeded to seek confirmation from other observers. He called Mike Feast, director of the South African Astronomical Observatory, to see whether anyone there had had any luck. As it turned out, all sites had been cloudy except Cape Town,

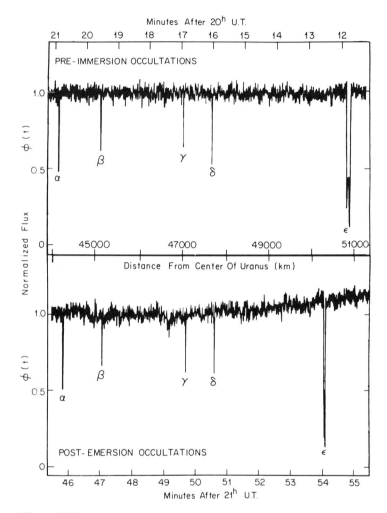

Figure 3.6

Rings rediscovered. The five large narrow dips in our signal that occurred before Uranus occulted the star agreed nearly perfectly with the five that occurred after the planet occultation when the two sets were aligned according to their projected distances from Uranus in its equatorial plane. This symmetry strongly implied that the brief occultations were caused by five narrow rings, rather than a belt of small satellites that we had originally considered. [Reprinted, by permission, from J. L. Elliot, E. Dunham, and D. Mink, "The Rings of Uranus," *Nature* 267 (1977), 328, copyright 1977 by Macmillan Journals Limited] where Joe Churms had recorded the occultation with the 13-inch refractor. Brian had calculated, from the times of the events that I had given him for our data, when the corresponding events should have occurred at Cape Town. To Brian's amazement, Churms had recorded the same events, along with several others that turned out to have been caused by bursts of electrons in his detector.

By now, Bob had returned from Perth, and Brian could check the event times in his data against ours. Three of his five matched perfectly with three of our five. (This discrepancy was resolved shortly thereafter.) Brian called to tell me of his inquiries and remarked, "I think you might have something with this ring business." The IAU circular went out with no more delay.

4

The Strangeness of Narrow Rings

The announcement in the IAU circulars started the bandwagon rolling. The world accepted with few questions the existence of rings that a few days before seemed theoretically impossible, if not esthetically displeasing. However, their existence did not make it any clearer how narrow rings could avoid spreading into one large disk. Dynamicists jumped on that problem, but they would need a careful analysis of the March occultation data and, in all likelihood, more occultation observations in order to solve the puzzle. As a temporary notation, I decided to denote the five definite rings by the first five Greek letters: α (alpha), β (beta), γ (gamma), δ (delta), and ϵ (epsilon). I intended to switch to the Roman letter notation used for Saturn's rings after we had learned exactly what structures composed the Uranian ring system, since we saw features other than these five rings in our data. It was not clear whether these were complete rings, fragments of rings, or something else that we had not thought of yet.

Altogether, weather conditions permitted eight groups to observe the occultation at seven sites. The data from the different sites varied greatly in quality and quantity. As would be expected, the reactions by the different groups to their data ranged from not noticing anything unusual to proposing narrow rings, an interpretation that only we, the KAO group, could make without depending on the observations of others.

While clouds and rain shut out several other groups at various sites in South Africa, Joe Churms, assisted by P. J. Booth, was observing the occultation on the 13-inch McClean telescope, a venerable instrument that had been in operation at the Royal Observatory since 1890. Although the main observing force had set up at Sutherland, Joe thought something should be done at Cape Town, and he was familiar with the McClean telescope and its photometric equipment. Uranus was low on the horizon when the disappearance of the star behind Uranus caused a 1-centimeter drop in the signal on the chart recorder. After 32 minutes, the signal rose back to its original level, but in the telescope the star was either invisible in the glare from Uranus or had not really reappeared. From his experience working with David Evans on an Antares occultation by the Moon in the early 1950s, he had learned not to stop recording data until he was sure that the occultation was over—even though it might *appear* to be over on the chart recorder. In order to separate visually the stellar image from the image of Uranus, he went to the 18-inch telescope for a sharper view and left Booth to carry on with the observations at the McClean telescope. In his absence, the chart pen dropped 1 centimeter for a few seconds and then returned to its original level. Miss Booth checked the position of the star in the guide telescope and found the star still centered on the crosswire. Joe found the star visibly separated from Uranus in the 18-inch telescope, so when he returned he knew it was safe to terminate the data recording. But what caused the dip? This remained a puzzle. Joe went to Sutherland, leaving his chart record with the director of the observatory, Mike Feast.

Observations were made in India by J. C. Bhattacharyya and K. Kuppuswamy at Kavalur and by H. S. Mahra and S. K. Gupta at Naini Tal. The Kavalur group noticed the dip caused by the ε ring and promptly reported it to the IAU circulars, thinking they had discovered a new satellite. After this they noted a 5 percent drop in their signal, which they interpreted as a grazing occultation by Uranus. They stuck to their satellite interpretation for quite some time, so that the result appeared in a paper in the British journal *Nature*, following papers by my group and the Lowell group that announced the discovery of the rings. Later they wrote a paper noting 19 dips in their data supposedly caused by rings—many, many more than anyone else had seen. To date they have no corroboration of their multitude of rings. Mahra and Gupta, the other Indian group, sent their event times to Marsden in April, apparently finding the ring occultations in their data after learning from others that they should be there.

Most of the Arizona teams had bad luck with the weather, except Ben Zellner at Perth. He recorded several ring occultations, but had packed up and left for home without realizing what he had. Eventually, he found them when the Lowell group hinted that he should "look it over carefully." In Japan, Koichiro Tomita made observations at the Dodaira Station of the Tokyo Observatory. Examining his data after learning of the ring discovery, he found 14 events. However, as he described in his paper in the *Tokyo Astronomical Bulletin*, none of these corresponded to the rings we had reported "for an unknown reason."

A few weeks later, the popular magazine *New Scientist* reported that a group headed by Dao Chen at the Purple Mountain Observatory in China had independently discovered the rings; the following year, they described their observations in a paper in the *Chinese Astronomical Journal*. Their data clearly show first contact by the ϵ ring, but the existence of events corresponding to the other rings had to be extracted by statistical methods. These methods also gave significance to several other events that do not correspond to any of the known rings.

Interestingly, all groups seemed to follow a similar line of reasoning, the differences being how far they got. The first step was to recognize dips in the data. Next was to eliminate an isolated cloud or poor tracking as the cause of the dips and then attribute them to a "satellite" or "satellites." Then came the "belt of satellites" idea, followed by the narrow ring conclusion. The KAO observations allowed us to work rapidly through this chain of reasoning to the narrow-ring conclusion and to be confident enough in our results to report them publicly. First, we could definitely eliminate the possibility of clouds by using Pete Kuhn's water vapor monitor and the possibility of tracking problems by watching the image on our television monitor. In addition, we had the only multicolor observations, so that we could be sure that only the star was being occulted. A nearby asteroid, for example, could have caused dips in our signal by occulting, or partially occulting, the planet and the star together. Having formulated the "satellite belt" hypothesis before Uranus occulted the star, we could check our conclusion by revising our observing plan on the spot to catch the "satellite belt" on the other side of Uranus. That gave us the only observations showing more than one ring occultation on both sides of the planet, which allowed us to demonstrate that narrow rings caused the brief occultations. Finally, the large investment in our expedition and the more public nature of KAO observations prompted us to report our results as soon as we felt confident of them.

In our paper to *Nature*, "The Rings of Uranus," we presented this conclusion: "At least five rings encircle the planet Uranus—as indicated by five brief occultations of the star SAO 158687 that occurred both before and after its occultation by Uranus on March 10, 1977." Looking ahead to exploiting further the occultation technique to observe rings, we suggested that "future occultation predictions for Jupiter, Uranus, and Neptune should include information about the apparent path of the star relative to the satellite orbit planes of these planets. The possibility that Jupiter has rings composed of 'rocky' bodies cannot at present be excluded. Saturn's rings may contain narrow components, of much higher optical depth than the average, akin to the rings of Uranus."

Had anyone detected the Uranian rings before without ever recognizing them? Herschel reported a ring around Uranus that he later retracted; he could not have seen the rings as they are now. The most sensitive of the observations that might have shown the rings was a long-exposure photograph taken by Bill Sinton of the University of Hawaii in 1972. Searching for inner satellites of Uranus, he exposed the photograph at wavelengths where the methane of Uranus itself absorbs red light strongly and the planet is thus faint. After the discovery of the rings, Bill reexamined his photograph, but could not see any rings. Knowing their approximate dimensions, he calculated that the rings must reflect no more than about 5 percent of the light from the Sun or they would have shown up on his photograph. He concluded, therefore, that the particles could not be coated with ice, which would reflect nearly all of the light. Sinton reported his findings in a paper in *Science* entitled "Uranus: The Rings Are Black." More recent measurements by Phil Nicholson of Cornell University and his collaborators show that the reflectance of the rings, their albedo, falls between about 2 and 3 percent for infrared radiation; similar values probably hold for the wavelengths that Sinton was using. Hence he could only have seen them in his 1972 photograph had the rings had three times their actual area or reflected three times as much light.

Another way the rings might have been discovered earlier was in photographs of Uranus taken through a 36-inch balloon-borne telescope known as Stratoscope. Flown by Bob Danielson and his group at Princeton, Stratoscope took photographs that were near the theoretical resolution limit imposed on the telescope by diffraction, rather than the degraded resolution imposed by the Earth's atmosphere on ground-based photographs. Disappointingly, the increased resolution of Stratoscope showed no detail on Uranus, although Jovian-type cloud patterns would have shown up. After the ring discovery, Giuseppe Colombo noted that the shadows of the rings might be visible on the photographs. Some people see them there and others do not.

Had anyone predicted the rings on theoretical grounds? What mainstream theories there were concerning the formation of rings attempted to explain the presence of Saturnian rings and their absence around other planets. No one had published a serious prediction of the Uranian rings in a major scientific journal. However, two claims to prediction were staked. The first that came to my attention was a paper on Saturn's rings by an amateur astronomer, Judd Boynton, which he had presented at an amateur conference in 1975. It contained the sentence, "With sharper examination, Uranus should likewise display rings." The other claim was by a former student of Nobel Laureate Hannes Alfvén, Bibhas R. De, who had submitted a paper to *Icarus* entitled "On the Possibility of the Existence of a Ring of Uranus" in 1972. The editor rejected it on the basis of referees' recommendations. After the discovery, De submitted it to *Icarus* again, and it was rejected again. Eventually the journal *The Moon and Planets* published it. Astronomers did not give much credit to either of these claims because each was based on a theory that appeared incorrect. Fair or not, to be right for the wrong reason does not hold much weight in scientific circles, especially regarding a question that has only two possible answers: yes or no.

Although the origin of the rings posed an interesting question, the more immediate problem was to explain the double-barreled surprise of narrow ringlets and sharp edges. Each of the three major contributors to ring theory—Laplace, Maxwell, and Jeffreys—had discussed the concept of narrow rings, but only in the context of their role as building blocks of a broad ring. Their goal, after all, was to account for observations of Saturn's broad rings, not isolated narrow ringlets, no observations of which had been corroborated. Laplace's quest for mechanical stability of a solid, broad ring led him to compose it of many narrow rings, each one constructed to be mechanically stable. In his monumental treatise *Celestial Mechanics* he states that "each ring is a solid body, whose center of figure coincides nearly with the center of Saturn; but that the center of gravity of the ring can be, and must be, at some distance of the center of Saturn."

In his Adams Prize essay, Maxwell showed that a stable solid ring would be so asymmetric in its mass distribution that it would conflict with the observed symmetry of the ring. On the other hand, the stability of a particulate ring was readily achieved. In reaching this conclusion, he considered the stability of broad rings, individual narrow rings, and systems of many narrow rings. On the subject of whether narrow rings could be broken up by wave motion of the entire ring, he concluded,

Passing to the case of narrow rings, we should find a somewhat larger maximum density, but we should still find that very short waves produce forces in the direction opposite to the displacement, and that therefore, as already explained, these short undulations would increase in magnitude without being propagated along the ring, till they had broken up the fluid filament into drops. These drops may or may not fulfil the condition formerly given for the stability of a ring of equal satellites. If they fulfil it, they will move as a permanent ring. If they do not, short waves will arise and be propagated among the satellites, with ever increasing magnitude, till a sufficient number of drops have been brought into collision so as to unite and form a smaller number of larger drops, which may be capable of revolving as a permanent ring.

The main thrust of Jeffreys's work was to investigate the role of collisions and show that an initial spherical cloud of particles orbiting a planet would rapidly become an equatorial disk and spread until a nearly collisionless monolayer evolved. However, his thought also included "narrow rings" as building blocks. In a 1916 paper on Saturn's rings in the *Monthly Notices*, Jeffreys wrote that "the constituent particles must then be distributed in narrow circular strips, each strip being practically a broken solid ring, and perhaps even undergoing something not very different from solid friction from the two on each side of it, which must be moving with slightly different velocities." These early ideas were not wrong, just not complete. Later theorists would build on this work.

By the middle of this century it did at least seem clear to astronomers why rings were flat and broad. Collisions require that a cloud of particles about a planet evolve into a broad, equatorial ring—a few particles thick at most. The strongest of Kirkwood's resonances might sculpt a single disk into a few broad rings, but the driving force of collisions would largely have its way. But now the occultation data had given us a glimpse of the Uranian rings that had over a thousand times better resolution than our best data on Saturn's rings, revealing narrow rings with sharp edges. Ring dynamics must be more subtle than what could be explained by collisions and a few major resonances.

Before considering the modern refinements to ring theory, let us trace Jeffreys's reasoning to see why rings should be flat—really flat. For a billiard ball of a planet—one perfectly spherical—each ring particle would move in an elliptical orbit, one focus of which is at the center of the planet. Actually, an elliptical orbit combines two kinds of motion, a perfectly circular motion (although not at a uniform rate) and an oscillation that first carries the particle outward to its farthest point, its apoapse, and then inward to its nearest point, its periapse. Because the time for a complete circular motion and the time for one radial oscillation are equal, the particle ends up at exactly the same point that it started; then it traces the very same ellipse again. It will never trace anything but that ellipse.

But then, real planets are not billiard balls. They all rotate, which raises a bulge at their equators and squashes their poles. A ring particle does not feel an oblate planet's gravity as if its mass were concentrated

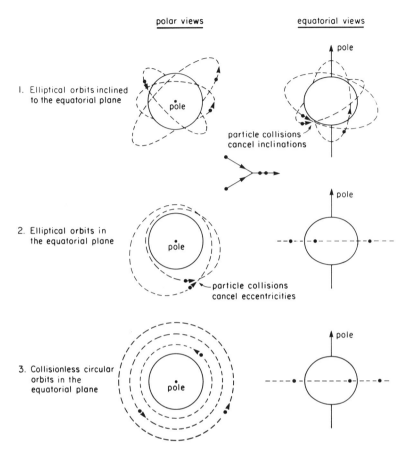

Figure 4.1
A cloud goes flat. If we begin with a "cloud" of particles around a planet, each would be revolving in its own orbit. Particle collisions would dissipate energy and cause the orbits of all particles to evolve rapidly to the same plane—the equatorial plane for a rotating planet. Once in the same plane collisions would continue to act, causing the particles to spread out and decreasing their orbital eccentricities until reaching the final (perhaps ideal) state of circular, collisionless orbits.

at its center, as would be the case for a spherical planet. Instead, the particle feels the combined effect of a sphere's point mass and the planet's "spare tire" around its equator. This combined gravity field causes the particle to move in an orbit that is almost elliptical, but not quite. As often happens, the "not quite" makes things more interesting. The particle still has two kinds of motion, circular and oscillatory, but the planet's oblateness has made the period of the radial oscillation longer than that of the circular motion. The particle must now complete a 360 degree circuit of the planet and then some before it reaches periapse again. It still follows a close approximation to an elliptical orbit, but the ellipse is steadily swinging around one focus. This continuous advance of its periapse, called precession, is faster for particles orbiting closer to the planet.

A similar effect occurs for "inclined" orbits—those not in the equatorial plane of the planet. In addition to the circular motion and the radial oscillation, the particle also has a vertical oscillation. For a spherical planet, the periods of these three motions would be equal, and the particle would follow the same inclined ellipse again and again. However, the oblateness of the planet also affects the period of vertical oscillation, making it shorter (for orbits inclined <63°) than that of the circular motion. The points where the plane of the orbit intersects the equatorial plane, called nodes, continually move backward (regress) around the equatorial plane. The closer the orbit to the planet, the faster the nodes regress.

If an initial cloud of particles surrounded a planet, either left over from the formation of the planet itself or derived from the tidal breakup of a moon, each particle would be following its respective inclined, elliptical orbit. Without collisions, each particle would continue in its respective orbit, which would precess and regress because of the planet's oblateness. But the precession and regression are bound to lead to collisions, even if the original orbits did not intersect. A collision of two neighboring particles in nearly identical inclined orbits would not change things much, since their small relative velocity would produce the gentlest of collisions. Since the orbit of the one closer to the planet would regress faster, however, the particles would eventually be crossing the equatorial plane in opposite directions. If the particles had equal mass, a head-on collision would cancel the momentum of each particle perpendicular to the equatorial plane, confining the motion of each to the equatorial plane. Although this is the ideal case, the principle illustrates how different nodal regression rates cause the collapse of particles into the equatorial plane of a planet in tens of thousands of years—a wink of an eye compared with the age of the solar system, which is about 4.6 billion years. In the absence of precession, for a planet that has absolutely no equatorial bulge, inelastic particle collisions would still cause the cloud to collapse to a disk, but at a slower rate.

Once the particles are orbiting in the equatorial plane, collisions again would cause further changes in their orbits. In a collision, both total energy and angular momentum are conserved. Some of the kinetic energy of the orbit would heat and deform the colliding ring particles, and some would set them spinning. However, there is nowhere else for the orbital angular momentum to go, except for a small amount that goes into spinning the particles. The result is that the eccentricity of the orbits decreases to zero (that is, they change from ellipses to circles) and the particles spread apart until collisions no longer occur. This is Jeffreys's monolayer of particles in circular orbits, the final evolved state of Saturn's rings. "It appears unlikely," he wrote, "that any measurable thickness [of Saturn's rings] will ever be found." Since collisions cause ring particles to spread until neighbors no longer collide, narrow rings would be difficult to understand — sharp edges doubly difficult to understand. What about the sharp-edged Cassini division? Theorists assumed that it probably involved the strongest resonance with the satellite Mimas, although they did not understand exactly how a resonance would clear a gap or how wide the gap should be. Several explanations were possible. The latest idea was put forth by Peter Goldreich and Scott Tremaine, who proposed that a spiral density wave created at the resonance would clear the gap.

Since nature had produced narrow rings, there must be a physical explanation. At first, the most promising cause seemed to be some combination of resonances with the five known satellites of Uranus because the position of exact resonance would be narrow like the rings. Although several people were thinking about how resonances could produce narrow rings, Stan Dermott and Tommy Gold at Cornell wrote the first paper proposing this mechanism. Tommy had brought Stan to Cornell while I was away for the occultation observations to work on resonance phenomena in the solar system. They showed that the positions of some of the rings closely matched the positions of some resonances. However, subsequent papers in *Nature*, one by Kaare Aksnes and the other, entitled "The Revenge of Tiny Miranda," by Peter Goldreich and his graduate student Phil Nicholson, exposed problems with the resonance idea: the proposed resonances were not strong enough to do the job; rings produced by these resonances would be only tens of meters wide, not the kilometer widths observed; and the strongest resonances, those with the satellites Miranda and Ariel, had been neglected. The simple coincidence of a resonance and a ring feature was not a sufficient explanation.

In the meantime, Goldreich and Tremaine had become intrigued enough with the narrow ring problem to try to make resonances fit the rings. Scott left Caltech (California Institute of Technology) for Cambridge, England, in August 1977, but kept corresponding with Peter about the problem. During the next year, their attempts to explain narrow rings with resonances did not work, so they tried a new approach. They pursued the counterintuitive idea that small satellites near the rings—through the attractive force of gravity— were "repelling" or "shepherding" the ring particles by collision-mediated inhibition of spreading. From their previous work with a spiral density wave that propagates through the Cassini division, they had the necessary mathematics for the force that a small satellite would exert on the ring particles. They met at a scientific meeting in Cortina, Italy, during July 1978, where, at the expense of not attending the talks, they wrote most of their *Nature* paper, "Toward a Theory of the Uranian Rings," describing the shepherd mechanism. When they finished, they noticed that the mathematics of the problem turned out to be quite similar to that used by Bill Julian and Alar Toomre of MIT, over a decade before, for describing the motions of stars in galaxies. In the mathematics, stars and ring particles play analogous roles.

A narrow ring in the Goldreich and Tremaine theory would require two shepherd satellites, one inside the ring and the other outside. The outer satellite's gravity raises a local bulge in the outer edge of the narrow ring. The particles in the bulge, moving faster in their orbit than the satellite, carry the bulge ahead of the satellite. This pulls the satellite in the direction of its motion, which adds angular momentum to the satellite and causes its orbit to expand. The ring particles lose angular momentum, causing their orbits to contract and giving the appearance that the satellite is "repelling" the ring particles. Collisions between ring particles destroy the bulge before that part of the ring gets back around to encounter the satellite again. An analogous reaction occurs between the ring particles and the inner satellite, except that the inner satellite is moving faster than the ring particles, so that it is forced to a smaller orbit and the ring particles to a larger one. With the outer satellite pushing the ring particles inward and the inner one pushing them outward, the net result is a narrow ring.

Another method for confining a narrow ring, proposed by Dermott and his colleagues, would require only one satellite per ring. In their view, a single satellite embedded within the ring keeps the particles from spreading through collisions. The satellite would also serve as the source of the ring particles. As the surface of the small satellite is eroded, perhaps by collisions with meteors or stray ring particles, the particles chipped off become trapped in horseshoe-shaped orbits that have a small gap near the satellite.

As theorists attempted to apply Newton's laws of gravity to narrow rings, we were finding out more about them by analyzing our observations. The interpretation of the five main rings (α, β, γ, δ, ϵ)

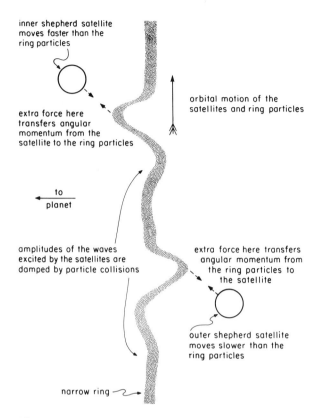

inner shepherd satellite
moves faster than the
ring particles

extra force here
transfers angular
momentum from the
satellite to the ring particles

orbital motion of the
satellites and ring particles

to
planet

amplitudes of the waves
excited by the satellites are
damped by particle collisions

extra force here transfers
angular momentum from
the ring particles to
the satellite

outer shepherd satellite
moves slower than the
ring particles

narrow ring

Figure 4.2
How do shepherds keep a ring from spreading? At first narrow rings defied
understanding, since we would expect ring particles to collide and spread out instead of bunching up. Peter Goldreich and Scott Tremaine proposed that a narrow ring could be maintained by two satellites, one orbiting inside and the other
outside the ring. We can understand the basic principle of this mechanism by
thinking of it in terms of tidal bulges and the resulting forces between the bulges
and the satellites. The inner satellite raises a small bump in the ring, which lags
behind the satellite because the orbital velocity of the ring particles is less than
that of the inner satellite. The force between the bulge and the satellite adds angular momentum to the ring particle orbits at the expense of the satellite's orbital angular momentum. This causes the ring particle orbits to expand and the
satellite orbit to contract, giving the impression of gravitational "repulsion." Particle collisions cause the bump to disappear before the same particles encounter
the satellite again. The same type of interaction occurs with the outer satellite,
which results in a contraction of the ring particle orbits. The result of inward and
outward forces from the shepherd satellites is a narrow ring.

seemed secure. Their occultation dips appeared where expected in all data sets that were not too noisy. Moreover, the broadest ping, ϵ, was wide enough to show the same wavy structure in the occultation profiles from both Perth and tD3 KAO. Since these stations probed the ring at points 3,000 kilometers apart, the ring mues maintain its structure well around its circumference.

The initial discrepancy between the times of two of Bob's dips and two of ours was resolved when we discovered that he had briefly interrupted his observations for calibration measurements when the α and δ rings occulted SAO 158687. His other two dips, denoted "4 and 5" in the Lowell *Nature* paper, would correspond to rings interior to the α ring. Although our data showed some shallow dips in this region, they did not seem to correspond to the same "rings" as calculated from the Lowell data. Depending on which of our three data channels we plotted, we could see from one to three dips in the signal. Were they fragments of rings? We dubbed these the "inner rings," but did not understand what they really were.

The next confirmed ring was a broad one—about 65 kilometers wide—that I found between the β and γ rings during the summer of 1977. It appeared on both sides of the planet, but caused only a 5 percent drop in the signal. So I asked Bob to check his data for the feature. Sure enough, the shallow, broad feature appeared exactly in the correct location. I christened it the η (eta) ring.

After combining all available data from the March 1977 occultation, Ted Dunham, Larry Wasserman, and I calculated the orbits for the rings. We found that some of the rings had different apparent radii on different sides of the planet, which could mean that they were elliptical or that they did not lie in the same plane with the other rings. We could not decide between these possibilities, since these data referred only to one orientation of the rings. To resolve this geometrical ambiguity, we would have to see the rings from a different angle, which would be possible after Uranus moved along in its orbit. In working out our method for calculating the ring orbits, we were continually impressed by the small effects that we had to include in our calculation to achieve a precision of 1 kilometer in our answer. For example, Einstein's theory of general relativity holds that light "bends" when passing near a mass; Sir Arthur Eddington's measurement of the bending of starlight by the Sun during the 1919 solar eclipse was one of the classic tests of the theory. Uranus, the ratio of whose mass to the Sun's is 44 : 1,000,000, bends starlight grazing its edge through an angle of 0.002 arc-second—but that proved significant, since this angle subtends a distance of 25 kilometers over the distance between

Uranus and the Earth. Fixing the center of Uranus at the point determined by our ring orbits, Ted and I were able to achieve one of our initial goals of observing the occultation by Uranus. We calculated an approximate radius and oblateness for the planet, using the chords measured from the KAO data and from Cape Town.

By the summer of 1977, we knew that Uranus had six rings, some elliptical (eccentric) or inclined, with some probable ring "fragments" inside the α ring. The dark, narrow rings with broad gaps between them appeared to be a "negative" of Saturn's bright, broad rings with narrow divisions, although Saturn's divisions are not nearly so narrow as the Uranian rings. We had learned much within a short time, but no more occultations had been predicted; nor were there likely to be any predicted for quite some time. Possibly one of the Voyager spacecraft could reach Uranus a decade later, but NASA had not yet decided to attempt this. A Uranus encounter would require a target point at Saturn that conflicted with a close approach to Titan—Huygens's 1655 satellite discovery that gave him the clue to interpreting Saturn's enigmatic companions as a broad, thin ring. Titan had attracted special interest in 1944, when Kuiper reported his discovery of its atmosphere. Close-range photography and other measurements by the two Voyager spacecraft could prove interesting indeed. Although NASA eventually decided to send *Voyager 1* close to Titan and *Voyager 2* on a trajectory leading to Uranus and Neptune, the prospect of more observations of the Uranian rings looked bleak at summer's end in 1977.

With a growing interest in more observations of the rings, Brian Marsden approached Arnold Klemola of Lick Observatory with the idea of photographically searching the star fields through which Uranus would pass in the near future. The photographic plates could reach much fainter stars than the approximately ninth-magnitude limit of the SAO catalog, he reasoned, so that some useful occultations might be predicted. They published a list of 12 stars that the rings would occult during the next 3 years. The prospect of more occultations was exciting, but the faintness of the stars left the value of observing these occultations in doubt. In December 1977, Bob Millis observed an occultation by the brightest star on their list, which was nearly six times fainter than the star occulted in March. He succeeded in detecting single occultations by four of the six rings, with a possible detection of the α ring. The width and radius of the ε ring were intermediate between the two cuts of the discovery observations. This result supported the idea that the ε ring is a continuous ring, rather than the fragmented model that Dermott and Gold had proposed. Disappointingly, the quality of Bob's observations showed that we could not expect to learn much about the rings from occultations by fainter stars on Klemola and Marsden's list.

At Caltech, Peter Goldreich was eager for more information about the rings to further his dynamical studies. After he saw the list of stars published by Klemola and Marsden, he talked to his colleagues— Gerry Neugebauer, Keith Matthews, and Jay Elias—and found that, although the stars might be too faint to yield useful results at visible wavelengths, observations in the infrared at a wavelength of 2.2 microns looked promising. Here, the absorption by methane in the Uranian atmosphere is much stronger than in the bands at visible wavelengths that we had exploited, so that the planet reflects only a small fraction of the incident sunlight. At these wavelengths, Uranus is even fainter than the rings themselves. To boot, most stars are relatively brighter than the Sun at 2.2 microns compared with visible wavelengths. The next opportunities to try observations in the infrared came in the spring of 1978. After a few attempts with the 2.5-meter telescope at Las Campanas Observatory in Chile, Eric Persson successfully recorded the passage behind the rings of the fifth star in Klemola and Marsden's list on 10 April 1978.

Starting with the circularity of the γ and δ rings that we had es- tablished with the March 1977 occultation, Phil Nicholson calculated the radii within the Uranian equatorial plane corresponding to the occultation dips in Persson's data. The five main rings appeared in their correct locations, the radii of the ε ring falling within the extreme values of 51,050 and 51,700 kilometers that we had found earlier. However, the width of the η ring appeared to be only a few kilometers, in contrast with the 65 kilometers that we had found from the 1977 data, and its radius was about 30 kilometers less than we had calculated. In the region of the "inner rings," they found six occultations, three on either side of the planet, whose radii matched closely enough in pairs to suggest the presence of three rings rather than unmatched fragments. Phil called me with these results, and I started thinking about how we could resolve these discrepancies.

We checked our calculations and found an error in one of Mink's programs that plotted our data. The radius scale previously calculated for our "fragments" had been wrong; our corrected results now agreed with the radii from the Lowell and Caltech data. In naming these rings, we changed notation from the Greek letter system because some groups had already used subsequent letters to denote events that turned out not to be rings. Since two of the rings corresponded to events 4 and 5 that the Lowell group reported in their original *Nature* paper, we called the three inner rings 4, 5, and 6, with 6 being the

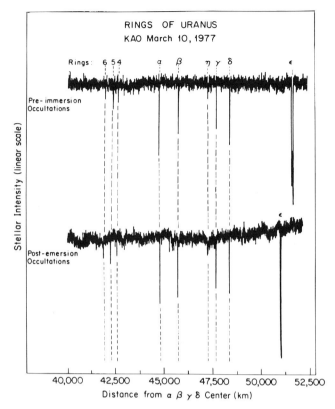

Figure 4.3
Nine narrow rings of Uranus. By adding signals from the three data channels of
the photometer that we used on the KAO, we can easily see matching occultations by all nine rings. Occultations by the ε ring occur at different distances
from the planet because this ring is markedly elliptical. [Reprinted, by permission, from James L. Elliot, "Stellar Occultation Studies of the Solar System," *Annual Review of Astronomy and Astrophysics* 17 (1979), 445, copyright 1979 by Annual
Reviews Inc.]

extension of the numerical notation. The discrepancies involving the
width and radius of the η ring remained.

Attention now focused on the ε ring. The Caltech group found that
the structure in their observations agreed in form with that found a
year earlier, but the width of the ring was different. Hence the ε ring
has a stable structure that "expands and contracts" in width, either
in time or around the ring or both. The different radii of the ε ring
implied that it could be inclined, elliptical, or some combination of
the two. Furthermore, if it were elliptical, it could either be centered
on the planet or have one of its foci coincide with the center of the

planet. For the latter possibility, it would be a precessing, eccentric ring. To avoid disruption by the different orbital precession rates of particles at different radii, the ring must precess as a unit, which implied that its width should vary linearly with its radius at the same location, its local radius. When Phil Nicholson plotted the widths of the five distinct occultation profiles available for the ε ring against their local radii, he found a straight line. Hence, the ε ring must be a precessing elliptical ring, whose width varied from 20 kilometers at periapse to 100 kilometers at apoapse. The oblateness of Uranus causes the precession. From the rate of the precession, the Caltech group calculated a value for oblateness of the mass distribution within the planet, which exerts the main influence on precession.

To maintain the uniform precession and avoid disruption, some mechanism must continually maintain equal precession rates for the inner particle orbits and the outer particle orbits. This would slow down the precession of the inner orbits and speed up the precession of the outer orbits, causing the entire ring to precess as a unit. Goldreich and Tremaine examined several mechanisms that might transfer the angular momentum and concluded that the most likely was the mutual gravitational attraction of the ring particles. However, collisions of the ring particles at periapse when the particles are most closely packed might also equalize the precession rates, as argued by Stan Dermott and his colleagues.

With the power of infrared occultation observations of the rings firmly established, Bill Liller suggested that we should observe more ring occultations ourselves by teaming up with infrared astronomers Jay Frogel and Jay Elias of the Cerro Tololo Interamerican Observatory in the Andes near La Serena, Chile. For the brighter stars in Klemola and Marsden's list remaining to be occulted, the 4-meter telescope there and its state-of-the-art infrared system would give us observations with better signal-to-noise ratio than had been yet obtained. Perhaps we could discover new rings; certainly we could learn more about the structure of the known rings, which might also allow us to calculate more complete orbits for the rings.

Since Uranus was in the southern sky and still moving south, the Southern Hemisphere observatories would see many more occultations with Uranus better placed in the sky than would those in the north. Accordingly, I arranged collaborations with other astonomers at major observatories in the Southern Hemisphere: Ian Glass at Sutherland, David Allen at Siding Spring, and Terry Jones at Mount Stromlo. In addition, Keith Matthews continued the Caltech observing program at Las Campanas. André Brahic and his graduate student Bruno Sicardy,

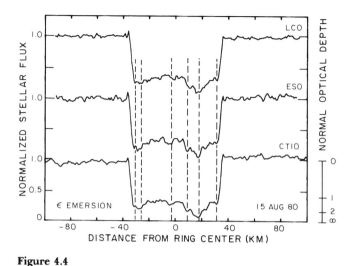

Figure 4.4

What would we see with a sharper view? These three graphs of the extinction of starlight by the ε ring were recorded during the August 1980 occultation of a star at three observatories in Chile: Cerro Tololo Interamerican Observatory (CTIO), Las Campanas Observatory (LCO), and the Europeon Southern Observatory (ESO). The dashed lines show bumps at the limit of spatial resolution that align for all three data sets. With improved resolution, such as can be obtained with Voyager stellar occultation observations, these features may prove to be very narrow, dense ringlets. [Reprinted, by permission, from J. L. Elliot, J. H. Elias, R. G. French, J. A. Frogel, W. Liller, K. Matthews, K. Meech , D. J. Mink, P. D. Nicholson, and B. Sicardy, "A Comparison of Uranian Ring Occultation Profiles from Three Observatories," *Icarus* 56 (1983), 202, copyright 1983 by Academic Press]

both of the University of Paris, and their collaborators began a program at the European Southern Observatory, near Cerro Tololo, and at other observatories in Europe.

The occultation on 15 August 1980, which was observed from three sites in Chile, brought more revelations about the rings. The signal-to-noise ratio of this event was the best yet. We saw the undulations in the signal caused by light diffracting around the edges of the γ ring, a "double dip" structure near the widest part of the α ring, and a possible tenuous, broad feature adjacent to the inner part of the δ ring. These data also solved the puzzle of the η ring: it was both broad *and* narrow! It had a broad component about 60 kilometers wide with a narrow component at its inner edge. The observations at visible wavelengths in March 1977 had been most sensitive to the broad component, while the infrared observations in April 1978 had been most sensitive to the narrow component. The signal-to-noise ratio for

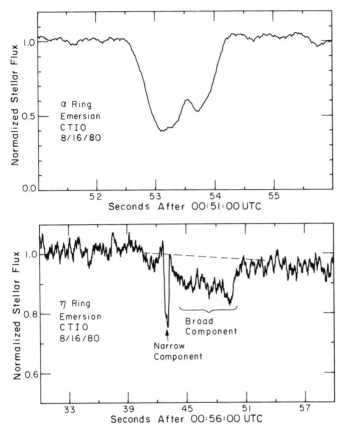

Figure 4.5

A pair of puzzles. The α ring (top) shows two dips separated by about 4 kilometers, the limit of spatial resolution, in this extinction graph recorded at Cerro Tololo Interamerican Observatory. Is it a "braided" ring? The same occultation event showed the η ring to be paired broad and narrow ring (bottom). [Reprinted, by permission, by J. L. Elliot, R. G. French, J. A. Frogel, J. H. Elias, D. J. Mink, and W. Liller, "Orbits of Nine Uranian Rings," *Astronomical Journal* 86 (1981), 444 copyright 1981 by American Institute of Physics for the American Astronomical Society] each of these events had not been great enough to show clearly the less prominent component in each case. Careful inspection of our KAO data revealed the narrow component at the inner edge of the broad component, now that we knew that it should be there. Our radius had differed from that obtained by the Caltech group because our radius referred to the middle of the broad component, while theirs referred to the middle of the narrow component. Another mystery solved.

This occultation also contained some unexplained dips in the data from the European Southern Observatory, which Brahic's group initially interpreted as possible additional rings. Their cause was most probably instrumental; certainly they could not be due to continuous rings around Uranus, since no corresponding dips occurred at Las Campanas or Cerro Tololo.

The August 1980 event also allowed a major improvement in the ring orbit model. Through the spring and early summer, Dick French and I had been developing a method for obtaining the orbital elements for each ring. This model would also give us information about Uranus: the precise position of its pole and measurements of how its mass distribution differs from spherical symmetry, which causes the ring precession. Using the ring as a bull's-eye target to locate the center of Uranus, we could calculate the radius and oblateness of Uranus from the planet's occultation times that we had obtained from several events, considerably improving upon our earlier result. Our value for the oblateness agreed reasonably well with the value obtained from a sophisticated reanalysis of the Stratoscope images by Fred Franklin and his colleagues. Finally, we could combine the value for the oblateness and our measurement of the mass distribution to calculate the rotation period of the planet: 15.5 $\pm$ 1.4 hours, assuming that Uranus is in pressure equilibrium. The rotation period of Uranus has been controversial in recent years, and different methods have produced values ranging between 12 and 22 hours. Our result, which agrees with some of the spectroscopic results, should turn out to be close to the correct value, since this method has been shown to give reliable results for Jupiter and Saturn.

Taking "pictures" of the rings proved to be difficult, but not impossible. Keith Matthews and his colleagues produced the first ring picture, scanning at two infrared wavelengths in order to remove the contribution of the planet light from the image. Of course, the resolution is about 10,000 times worse than what we had achieved with occultations, which causes the nine rings to appear as one. The ε ring comprises most of the image, since it has the majority of the surface

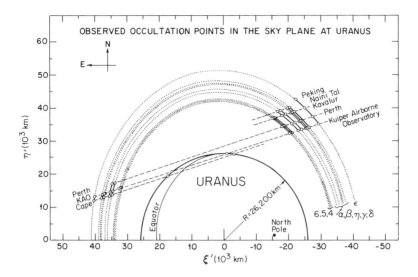

Figure 4.6
Connect the dots. This diagram shows the orientation of Uranus and its rings at the time of the discovery occultation in March 1977. The dots show the points of occultation as seen by each observatory. By combining precise timings from the occultations observed through the years, we have been able to establish the orbits of all nine rings, to determine the orientation of the Uranian polar axis, and to learn about the mass distribution within Uranus. [Reprinted, by permission, by J. L. Elliot, E. Dunham, L. H. Wasserman, R. L. Millis, and J. Churms, "The Radii of Uranian Rings α, β, γ, δ, ϵ, η, 4, 5, and 6 from Their Occultations of SAO 158687," *Astronomical Journal* 83 (1978), 980, copyright 1978 by American Institute of Physics for the American Astronomical Society]

area. A spectrum of the rings in the near infrared, obtained by Phil Nicholson and his colleagues, shows that the rings resemble carbon black, reflecting between 2 and 3 percent of the incident sunlight. There is no evidence in their spectrum for the absorption features of ice or frozen ammonia. The dark ring particles contrast markedly with the icy surfaces of the Uranian satellites and the even brighter, icy particles of the Saturnian rings.

Our glimpses of the Uranian rings through occultations gave us a picture of a ring system a thousand times sharper than we had ever seen before, revealing surprisingly narrow rings. The shepherd satellite model for confining the rings seemed plausible, but no one had yet detected a shepherd satellite—thought to have a radius of about 10 kilometers, or possibly even less. The shepherds might also account for some of the rings being ellipses, but the sharp edges of the rings remained a problem—as did the fascinating structure of the δ, α, η,

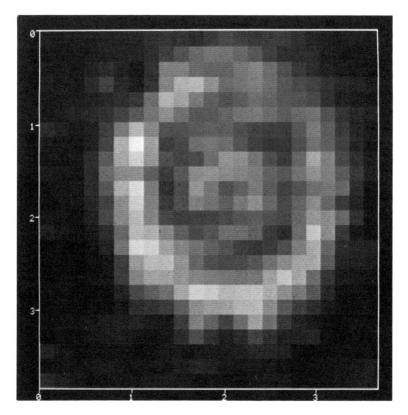

Figure 4.7
The Uranian rings seen with infrared eyes. This photograph of the Uranian rings was constructed from infrared scanning data obtained by Keith Matthews and his colleagues with the 5-meter telescope at Mount Palomar on 26 May 1983, at a wavelength of 2.2 microns. Infrared imaging of the rings has proved much more successful than imaging at visual wavelengths because the scattered light background from the planet itself is much less at infrared wavelengths, due to strong methane absorption in the Uranian atmosphere. Individual rings are not resolved, since turbulence in the Earth's atmosphere has limited the resolution of this photograph to about 20,000 kilometers, which corresponds to about 3 of the small squares ("pixels") in the figure. [Courtesy of K. Matthews, P. Nicholson, B. Soifer, and G. Neugebauer]

and ϵ rings. Self-gravity might well be the reason that the α, β, and ϵ rings precess as a unit, rather than smearing themselves out, but no one had any observations proving that this was the correct explanation. The Uranian rings seemed to be making some sense, although Saturn's rings had not prepared anyone for the shock of narrow rings. Were there more surprises to come?

5

Jupiter Joins the Ring Club

Saturn has rings, contrary to Galileo's expectations. Uranus has rings, contrary to twentieth-century expectations. Does Jupiter have rings too? Astronomers thought not, if they thought about it at all. They had the usual reasons—no one had seen any from Earth, and some arguments seemed to explain their absence.

Still, wondered Tobias Owen, should not someone make sure? The best opportunity ever for a close look at Jupiter was coming up, but no one else was showing much interest in the possibility of rings. Planetary rings were not one of Owen's particular specialties; he had studied planetary atmospheres under Kuiper using spectroscopy. When the two Voyager spacecraft whipped by Jupiter in 1979, his main interest would be the planet's atmosphere and its vibrantly colored clouds. Even so, as a member of the Voyager imaging team, he had taken it upon himself to propose a search for Jovian rings. All he had to do now was convince other Voyager scientists to give up some of their favorite observations to make room for the ring search. No, he had no hard evidence suggesting the existence of Jovian rings; but it was a mission of exploration, and exploration could be full of surprises, right? True, but there was more than enough for Voyager to study that scientists knew actually existed. Why waste precious time near the planet peering into what would probably be empty space when Voyager could be doing something of predictable value, such as analyzing the cloud particles in Jupiter's atmosphere? Yes, the discovery of the Uranian rings surprised everyone, but we already know a lot more about Jupiter than we knew about Uranus. Perhaps Jupiter's massive gravity could drag interplanetary debris into some sort of ring unlike Saturn's. How exactly could it do that? Owen's job was not going to be easy.

There was help at hand, though, if someone could uncover it and recognize it as such. The Voyagers would not be the first to Jupiter.

They had two scouts ahead of them—*Pioneers 10* and *11*. These simpler, lightweight probes were pathfinders that would search out the hazards along the route that the Voyagers must follow while making a variety of scientific observations. Of the 11 science experiments on each of the Pioneers, 6 measured radiation, in the form of electrons and protons, or the magnetic fields that might concentrate the radiation near Jupiter to lethal levels.

The first hazard, though, was the asteroid belt. Space jockies from Buck Rogers to Han Solo have dreaded such swarms of banging boulders. Astronomers, never having had such a close look, were less certain of what the first Pioneer would find. Pioneer investigators did extrapolate from the widely scattered, kilometer-size planetesimals that they could see from Earth to the more numerous pebble-size particles and dust that would pose the real threat to any spacecraft. They found that the asteroid belt was less intimidating than science-fiction heroes supposed, but was still potentially dangerous. They were not at all sure just how dangerous. The belt's dust might be so thin that Pioneer would have less than 1 chance in 100 million of being hit by one of the grains of rock whizzing around the Sun at more than 50,000 kilometers per hour. Or, it could suffer 10 damaging, perhaps lethal, hits before emerging from the asteroid belt. The world held its breath, or so the press had it, as *Pioneer 10* passed beyond Mars and entered the danger zone.

Nothing happened, which created new possibilities for Pioneer. William Kinard and his colleagues at NASA's Langley Research Center had expected that their meteoroid penetration detector would be useless after running the gauntlet of the asteroid belt. But one dust grain per cubic kilometer had not strained the detector at all, leaving it ready to probe the vicinity of Jupiter. The detector consisted of 234 gas-filled cells mounted on the back side (the leading surface) of Pioneer's big dish antenna. Sealed on its exposed surface by a 25- or 50-micron-thick wall, a cell records a hit when a micrometeoroid punctures it and the gas drains out, allowing a current to flow between its two electrodes. Once breached, a cell is lost for good and cannot record any more impacts. *Pioneer 10* had lost half of its cells to a malfunction in one of its two independent data channels 6 days after launch, and meteoroids punctured another 57 en route, leaving one-quarter of the original cells for the Jupiter encounter.

The vicinity of Jupiter, it turned out, is a very dusty place, 10 to 100 times dustier than the once formidable asteroid belt. Kinard and his group were not sure whether the 10-micron-size particles detected by *Pioneer 10* within 20 hours of closest encounter formed a cloud of particles orbiting Jupiter or were Sun-orbiting particles concentrated near the planet by its powerful gravity. "It is clear, however," they later wrote, "that the particles detected were not in a thin equatorial ring around Jupiter, similar to the rings around Saturn." The hits were scattered too far above and below the equatorial plane to have been in a ring. And besides, the detector went on recording hits at the same rate as before the encounter, suggesting that *Pioneer 10* had not lost many of its cells at once to a dense ring.

Then *Pioneer 11*, having cell walls twice as thick as *Pioneer 10*, recorded only four hits near Jupiter. Donald Humes, the group member who had been left the chore of final data analysis, concluded in 1976 "that the increase in the penetration fluxes observed when Pioneers 10 and 11 passed near Jupiter was the result of gravitational focusing of meteoroids having solar orbits." That was the only statistically supportable conclusion that Humes could reach, but what if the signal from a ring were buried in the noise from gravitationally focused meteoroids? Someone hypothesizing the existence of Jovian rings would look for hits recorded at each crossing of the equatorial plane. There could be no more than one on each of the three operating channels because each channel's event counter turns itself off for about 80 minutes after a hit to make sure that a false count is not registered from an incompletely drained cell. There would be no ring hits if the spacecraft did not happen to cross the equatorial plane within a ring.

It would not make for a strong statistical argument, but the meteoroid detectors on the Pioneers behaved exactly as they would if Jupiter had a broad ring. Within the precision of the impact timing, each of the three operating channels recorded an impact at the equatorial plane. *Pioneer 10*'s hit was only one of 10 hits scattered over almost 2 days, but the 2 hits as *Pioneer 11* pierced the imaginary plane stood all alone. The closest hits had come 7 hours earlier. Humes noticed the three impacts, but the statistics discouraged him. What good was a sample of three cases? He could never prove it was not chance. Besides, no one ever showed any interest; no one ever asked about the ring plane crossings.

The meteoroid penetration data lay fallow in the scientific literature. They needed a boost if they were to spark any productive speculation about rings around Jupiter. A Pioneer detector of far smaller particles had made just that sort of observation. One of the prime objectives of the Pioneers, in addition to scouting the asteroid belt, was the measurement of charged particle radiation—protons and electrons— that is spewed into interplanetary space by the Sun. Five different

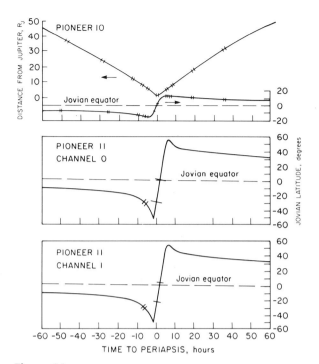

Figure 5.1

Hints of a Jovian ring. The Pioneer meteoroid detection experiment was not designed to probe the vicinity of planets in detail, but the results at Jupiter were
suggestive of rings. The meteoroid impacts recorded near Jupiter are shown here
plotted against time to periapse, or closest approach (time proceeds from left to
right), and Jovian latitude (any ring would lie at the Jovian equator). The top diagram appeared as figure 6 in D. H. Humes, J. M. Alvarez, R. L. O'Neal, and
W. H. Kinard, "The Interplanetary and Near-Jupiter Meteoroid Environments,"
Journal of Geophysical Research 79 (1974), 3677–3684 (copyright 1974 by American
Geophysical Union), and the lower two diagrams were later drawn by Humes but
are unpublished. *Pioneer 10* had only one operating channel (top), but *Pioneer 11*
had both channel 0 (middle) and channel 1 (bottom) operating. The bracketed intervals marking the impacts represent the 71 minutes deemed to be the maximum time required after a meteoroid pierces an individual gas-filled cell to drain
it completely and signal an impact. In practice, the drain time is only rarely
longer than about 2 seconds. Thus, the likely impact times are at the latest (farthest right) times bracketed. As plotted, all three channels have impacts recorded
near the equator, a particularly suspicious coincidence in the case of *Pioneer 11*'s
total of four recorded impacts. Because the experiment was not monitored con-
tinuously, a precise comparison of equator crossing times and impact times cannot be made. David Lozier and Jack Dyer of the Pioneer navigation team
provided a *Pioneer 11* equatorial crossing time of 2214 hours (ground received
time in the Pacific Standard Time zone). The impact on channel 0 occurred
sometime between 2200 hours and 59 seconds and 2213 hours and 23 seconds (a
difference of 1 minute), and that on channel 1 occurred between 2207 hours and
23 seconds and 2219 hours and 47 seconds (a period including the crossing). The
Pioneer 10 crossing at 1922 hours could coincide with the impact interval 1922
hours and 52 seconds to 1923 hours and 16 seconds.

experiments would be making the first measurements ever of charged particles near Jupiter.

Pioneer scientists had some idea of what to expect, having flown similar experiments on Earth satellites. There they found the Van Allen belt, radiation trapped near Earth by its magnetic field. Researchers assumed that they would find a Jovian magnetic field and a Jovian radiation belt because they had picked up radio waves from Jupiter almost surely produced by magnetically trapped electrons. As at Earth, charged particles near Jupiter would corkscrew along magnetic lines of force as they arced back and forth from one magnetic pole to the other. Jiggled by disturbances of the magnetic field, the charged particles would also slowly sift through the magnetic field lines toward the planet. This inward diffusion of electrons and protons would be like balls rolling downhill, gaining energy all the way. Approaching the planet, they would be so powerful and so densely packed as to be lethal, to man and perhaps machine.

That worried Pioneer scientists. The latest estimates of radiation intensities were running as much as 100 times higher than those used to design the Pioneers. But there were possible havens in the storm. Unlike Earth, Jupiter has an entourage of moons, some of which orbit well within the expected Jovian radiation belt. If they swept up charged particles the way Earth's moon seemed to sweep up solar electrons, then the Pioneers should be able to slip in close to the planet under the partial protection of the radiation shadows of the satellites.

Pioneer 10 did survive its 4 December 1973 encounter with Jupiter, while suffering negligible radiation damage. It found a huge Jovian magnetic field canted at an angle to the equatorial plane and a proportionally powerful radiation belt. And the inner satellites were indeed absorbing charged particles as the radiation mirrored back and forth through the Jovian equatorial plane. The spacecraft recorded the resulting peaks and valleys in radiation intensities while it plunged to within 132,000 kilometers of Jupiter's cloud tops (2.8 Jovian radii from the center of the planet). On 3 December 1974 *Pioneer 11* ventured to 1.6 Jovian radii, revealing the radiation shadows of Jupiter's moons Europa, Io, and little Amalthea, which orbits at 2.6 Jovian radii.

Pioneer 11 found something more. Inside the dip caused in all likelihood by Amalthea, there was a dip at about 1.8 to 1.9 Jovian radii, on both the inbound and outbound legs. No one was sure of their cause. In their first formal paper on the *Pioneer 11* results, submitted on 14 March 1975, Walker Fillius and his team at the University of California at San Diego concluded that the two outer dips recorded by their trapped radiation detector "may reasonably be attributed to particle absorption by Amalthea. However, since there are no more moons nearby, the other features require another explanation." Their best explanation at that point was that Jupiter's magnetic field might be more complicated than their simple model suggested. Perhaps the field folds over on itself allowing Pioneer to pierce the same field line twice, or the field could bend into the atmosphere, which would drain away charged particles.

In print, at least, magnetospheric specialists agreed. Mario Acuña and Norman Ness of NASA's Goddard Space Flight Center in Greenbelt, Maryland, submitted a paper to the *Journal of Geophysical Research* on 19 March that laid out three possible explanations for *Pioneer 11*'s dips in radiation intensity. The "most plausible" explanation, they agreed, was that their model was far simpler than the real magnetic field. A "second explanation" was losses to the atmosphere, again in agreement with Fillius's group. Their third possible explanation was a bit more speculative. "Finally," they wrote, "although we consider it remote, the possibility exists that the two minima are due to sweeping effects by an unknown satellite or ring of particles at approximately 1.83 R_j [R_j = 1 Jovian radius] not yet visually observed."

In private, Acuña and Ness took a bit stronger interest in possible undiscovered satellites or rings. Their paper would not appear in print for 9 months, and even then no planetary observer would ever see that bit of speculation, buried as it was in a study of particles and fields in a geophysical journal. So Acuña and Ness wrote a letter on 28 March to Bradford Smith, a Jupiter observer and head of the Voyager imaging team, a group that included Owen.

Dear Brad:
You may recall from our discussions at the DPS meeting in February that there is a possibility that the explanation of the complex radiation belt structure observed close to the planet Jupiter is only partially explained by our complex magnetic field. There are some features of the particle data which we are as yet unable to explain directly and the thought had occurred to us that perhaps there is either an undiscovered satellite close to the planet or perhaps a belt (or belts) of much smaller particles there. From our data analysis and interpretation, we would conclude that an absorber of charged particles may reside at a planetary centered distance of about 1.87 R_j in [sic] (or slightly closer) and in the equatorial plane. The evidence at present is not yet compelling but the exciting possibility suggest [sic] to us that we should appraise you of our present status and inquire as to your own thoughts on the matter.
May we hear from you at your early convenience? Best regards.

Smith responded on 7 April with some encouragement concerning a search for their possible inner satellite, but not much optimism about a ring search.

Dear Mario:

Thanks for your letter reminding me of the conversation at the DPS Meeting with you and Norman regarding the possibility of an undiscovered Jovian satellite within the orbit of JV [Amalthea].

Actually, I had not forgotten, and am presently looking into an observational technique which may improve the detectability of objects close to Jupiter. If it appears that we can make a significant improvement, we'll give it a try next October when Jupiter is near opposition.

There is always the possibility that rings of small particles exist, similar to the rings of Saturn, but of much lower number densities. The only opportunity to look for possible Jovian rings is at the time that the earth is passing through the equatorial plane of Jupiter. Unfortunately, the next opportunity will not occur until the summer of 1979, and at a time when Jupiter is quite close to solar conjunction. Even so, we'll give it a try.

I'll keep you informed of our plans to search for an inner satellite at approximately 1.8–1.9 R_j.

With best regards to you and Norm

Ness inquired again by letter in December, but received no reply. Smith had actually looked for an inner satellite but found nothing new.

Acuña and Ness were not the only ones trying to draw attention to the curious dips in the charged particle data. Fillius had seen their paper for the *Journal of Geophysical Research* in draft form and decided to give their satellite/ring idea a little more prominence. Seeking a broader perspective, he stood up in the audience at a May 1975 Jupiter meeting in Tucson, explained the proposal, and asked whether it was "preposterous or not." No response. Fillius assumed that the suggestion had not outraged anyone.

In his paper for the proceedings of the Tucson meeting, Fillius noted Acuña and Ness's speculation on the possible presence of something inside the orbit of Amalthea:

It would not have to be a single mass, for a ring of smaller particles could do the job as well. Indeed, since minima N2 and N3 [the unexplained dips] are inside the Roche limit, a particle ring is more likely. What with the similarities between Jupiter and Saturn, there seems to be no a priori reason against a dust ring near Jupiter, too. However, there are obvious questions which need to be investigated. Why has there been no optical detection of such a ring? What would be the gravitational effect on the other satellites and on the Pioneer spacecraft? Would one expect the Pioneer Meteoroid Detectors to detect the ring when they passed near it? This exciting hypothesis

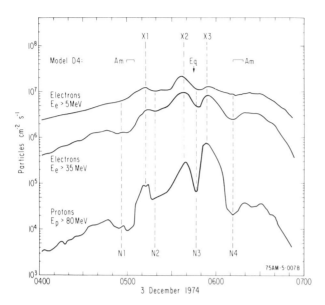

Figure 5.2

Charged particle astronomy at Jupiter. This plot of radiation intensity versus time shows the detection of one satellite and the possible trace of a second body or group of bodies. It was recorded by the trapped-radiation experiment aboard *Pioneer 11* as it swept below the planet within a distance of 1.6 Jovian radii. The dips marked N1 and N4 are probably due to absorption by Amalthea (Am). The dips marked N2 and N3 were of much less certain origin. Walker Fillius wrote in this figure's caption that "the multiple peak structure is unexplained, and could be accounted for by magnetic field anomalies, a dust ring, or some other cause." In the text, he argued at greater length that a ring could have caused it. [Reprinted, by permission, from W. Fillius, "The Trapped Radiation Belts of Jupiter," in *Jupiter*, edited by T. Gehrels (Tucson: University of Arizona Press, 1976), copyright 1976 by University of Arizona Press] has obvious problems, and it will take some time to sort out all of the possibilities and ramifications.

Interesting questions, but no one got around to answering them. Fillius could not convince Wing-Huen Ip, one of his graduate students, to pursue the problem. Humes had been at the Tuscon meeting, but does not recall hearing Fillius's comments. Smith seems to have put out of mind the possibility of a ring after his unsuccessful satellite search. Even Acuña and Ness eventually cooled on the idea. "We also suggest," they wrote in a 1977 paper with Juan Roederer, "that the peculiar behavior of particle counting rates seen by Pioneer 11 inside the orbit of Amalthea (Fillius et al., 1975) could be mainly due to field-geometric effects. . . . Only a study of detailed angular distribution data will reveal if this is indeed the case. . . ." They made no mention of unseen rings or satellites.

In October 1977, Owen had to make his final case for a Jovian ring search. The pathfinding Pioneers would play no role. Instead, his arguments were philosophical. Explorations could not always be reliable; the payoff could not always be certain. It entailed risk, but the possible loss was small and the potential gain large. Eventually, the working group allotting Voyager observing time agreed with Owen, as long as the observing time and the instructions to the onboard computer were short. Edward Stone, Voyager project scientist, concurred. In such cases, he reasoned, you do not let preconceived ideas keep you from making observations. Brad Smith would later comment that the search was not made "with any great expectation of a positive result but more for the purpose of providing a degree of completeness to Voyager's survey of the entire Jupiter system." Another view, expressed later by an imaging team member, was that some team members regarded the search as a way to prove once and for all what they already knew to be true—that there was no ring.

So Owen got his chance, such as it was. He and Edward Danielson of Caltech had started by asking for four exposures. Some were intended as bargaining chips, as was the custom on the Voyager teams, but the final allotment was a single, all-or-nothing, 11-minute exposure as Voyager crossed the equatorial plane. To fit the search into the allowed computer memory, Candice Hansen, a Voyager technician, whittled the camera control command down to about 8 words in a 2,500-word load of commands to the computer.

Owen's remaining problem was where to point the camera. Aiming at the middle of a Saturn-type ring, scaled to Jovian dimensions, seemed most reasonable. Conveniently enough, that targeting also

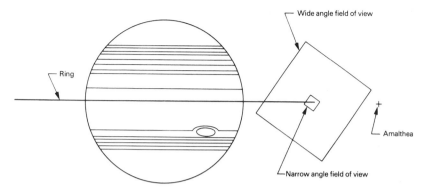

Figure 5.3

The lucky shot. The size of the wide-angle field of view and the relative proportions of Jupiter's rings set the area that *Voyager 1* would search with its single
exposure. Scattered light overexposed the wide-angle view, but the narrow-angle
camera, which had to point in the same direction, managed to catch the tip of
the edge-on ring. [Courtesy of NASA]

allowed the wide-angle camera to cover the equatorial plane from
near the planet almost out to Amalthea, the farthest any ring could
be expected to extend. Since the narrow-angle camera was rigidly
mounted beside the wide-angle camera and had to point at the middle
of that camera'a large field of view, it too would take an 11-minute
exposure.

When *Voyager 1* made its stab at detecting Jovian rings on 4 March
1979, Owen was huddling with other team members in their area at
the Jet Propulsion Laboratory (JPL) in Pasadena. An excited Candy
Hansen entered and greeted him with "Hey, Toby, we got something!"
She handed him a Polaroid print of the ring search image as it had
just appeared on a television monitor. "By God, there is something
on it!" was his first reaction. To his surprise, it was the narrow-angle
exposure, not the wide-angle search he had been depending on; that
one had been hopelessly overexposed by the heavy trapped radiation
and a miscalculation of the brightness of light scattered from Jupiter.

There was indeed something in the image, but what was it? There
was a stack of thin, fuzzy, straight lines and several brighter, jagged
lines resembling broken hairpins. At first, Owen took the fuzzy lines
to be rings, but then he began to have doubts. Others questioned the
ring interpretation too. Al Cook's calculations placed the images well
outside the equatorial plane. And Brad Smith, the one who would
have to retract publicly any erroneous announcements of a ring discovery, remembered all too well the premature announcement of a

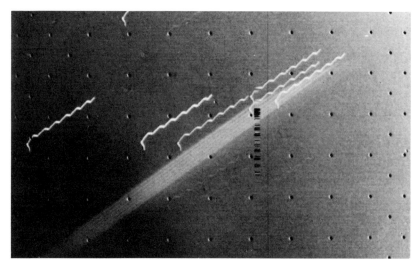

Figure 5.4
The Jovian ring discovery image (or images). The spacecraft's motion past Jupiter and its gentle rocking combined to produce this bizarre rendition of the Jovian ring seen edge on. The bright, squiggly lines are star trails, and the broad band is a stack of six ring images. The outer edge of the ring is at the upper right. The dark vertical lines near the center are caused by loss of data during reception at the tracking station. [Courtesy of NASA]

"moon" of Mercury discovered by the *Mariner 10* spacecraft—it turned out to be a star.

Owen and Danielson now had to prove their point. They had the advantage over Elliot of knowing what it probably was that they had detected. An added advantage was that, by chance, the narrow-angle camera had caught a cluster of stars in the same image. Once identified, their streaked images (the broken hairpins) provided a guide to the motion of the camera, both its sprint through the equatorial plane and its gentle nodding or rocking about its center of gravity. For each of the six kinks in a star trail there was a fuzzy image of the ring, which matched the calculated spacecraft rocking motion. And a re-calculation by Tom Duxbury of JPL put the ring images right in the equatorial plane.

After 3 days of groundwork by Owen and Danielson, Smith stood before the regular morning press conference. Hundreds of reporters from newspapers, magazines, radio, and television, cameramen, sound technicians, public relations people, science-fiction writers, and general hangers-on awaited their daily dose of discoveries and instant science. Smith's prime announcement today was the discovery of "a thin, flat ring of particles surrounding Jupiter. The discovery of the ring was unexpected in that the current theory that treats the long-term stability of planetary rings would not predict the existence of such a ring around Jupiter." Voyager's camera had just caught the outer edge of the ring 57,000 kilometers above the cloud tops. Because it had only been seen edge on, Smith could only say that the ring (or rings) was at least 9,000 kilometers wide and less than 30 kilometers thick. The ring's particles must be large, Smith suggested, because small ones would be quickly swept away. The absorption and reemission of solar radiation, called the Poynting-Robertson effect, and the drag of Jupiter's plasma, a tenuous gas of electrons and positively charged ions, would slow dust and pebble-size particles until they spiraled quickly toward the planet and a fiery end in the atmosphere.

After Smith had introduced Owen and Danielson as "two of those people who were very much involved," the questioning began. Was there any significance to the position of the outer edge of the rings? "As far as significance is concerned," Smith replied, "one always looks for some reason to explain why a ring has an outer edge, and it is close to a three-halves resonance with Amalthea. Amalthea is a very small body. Whether that three-halves resonance is adequate to provide that outer boundary is not yet clear."

What does the ring mean for the formation of the solar system? "Whether this material we're seeing in Jupiter's ring is primordial material that was left over at the time Jupiter was created or whether it is what's left of a satellite that unhappily got inside Roche's limit can't be answered yet," offered Smith. "It could be getting some material supplied to it from outside," added Owen, "that's another possibility." If that were so, wondered Smith, how could the outer edge be so sharp?

Why is it that this cannot be seen from Earth? Danielson had estimated that the ring would be so faint from Earth that it could conceivably be detected only by the largest telescope. "The ring is extremely thin [optically tenuous]," replied Smith, "so the only opportunity to see that ring is when it's edgewise. If we were up above the plane looking down on the ring, we probably could not see it. There are only certain opportunities that occur every 6 years to observe such a ring from the Earth . . . but it is not clear that even with the ability to observe at [Earth's] ring plane crossing that such a very, very weak feature could be seen against the glare of Jupiter. It's very close to the planet." The last two times that the Jovian ring would have been edge on as seen from Earth, Jupiter had been too close to the Sun. The next chance was 5 months away.

Will *Voyager 2* be able to follow up on this? Yes, said Owen, there will be opportunities at two ring plane crossings. "It's not now in the sequence," noted Smith. "The fact that we put it in Voyager 1 and didn't put it in Voyager 2 is an indication of what we thought the chances were of finding anything." "What some of us thought, anyway," added Owen, much to the amusement of the crowd.

Science News asked whether, with hindsight, anything of the ring in the *Pioneer 10* and *11* data could be seen. Owen took the question. Owen had made a phone call to Humes, who noted that *Pioneer 10* passed through the equatorial plane well outside of the ring at 2.88 Jovian radii, that the outer edge of the ring is at 1.8, and that *Pioneer 11* crossed inside of that at 1.67. "We don't know that it went through actual ring material because we don't know the width of the rings. However, Humes said that on Pioneer 11 there was an event, a meteoroid detection, at or near the plane crossing. Unfortunately, his timing on the instrument wasn't quite fine enough to say exactly when the event occurred."

Ed Stone offered another comment: "I've also looked at the Pioneer 11 encounter because it did penetrate inside the 1.8 R_j. I went back to look at the charged particle data, which was measured at that time, and in fact there was some very curious unexplained structure in the radiation intensity. I think Walker Fillius in writing a caption for one of the figures that showed the multiple peaking in the radiation intensity indicated that perhaps that might be due to a distorted magnetic field, and even suggested the possibility of a dust belt, but of course that was just a suggestion at the time. Clearly, there is a set of data there now that has to be looked at from a much different point of view than was the case when the Pioneer people looked at their data the first time." The whole story was not out yet, but it would soon be clear that Acuña and Ness had indeed looked at their data from the proper point of view. The 1.83 R_j position for their "remotely possible" satellite or ring was only 2,000 kilometers away from the edge of Voyager's ring.

The final question was whether it will ever be detectable by observing a stellar occultation. It would be very tough, replied Owen, because of the ring's tenuousness and the small target it presents. It is never inclined more than 3 degrees. "There is a chance to look for them in the infrared. That would be the best way to do it."

News of the Voyager discovery traveled a good deal farther and faster than word of the Pioneers' inklings of rings. Eric Becklin and Gareth Wynn-Williams, 4,200 meters up on the summit of Hawaii's Mauna Kea, read about the Jovian ring in the next day's *Honolulu*

Advertiser. They did not need Owen's prompting to realize that they could see the ring without bothering to send a spacecraft. They knew that the techniques commonly used in infrared astronomy to cut down glare and scattered light could have revealed the ring years before. Dropping their planned observing program, they spent two nights scanning the vicinity of Jupiter with the University of Hawaii's 2.2-meter infrared telescope. The key technique was the same one later observers of the Uranian rings would rely on—a 2.2-micron filter that darkened the planet, but not the rings, by a factor of 30. Although viewed as a confirmation of Voyager, their results—a distinct brightening at the equatorial plane near the planet—could not have convinced pre-Voyager observers. It would have intrigued them. But then, no one would have made the observation. It was too much of a long shot for a $5,000-a-night telescope; astronomers had more predictable science to do.

Now that the instant science of the press conference was over, the quick science could be done. Voyager scientists were too busy getting ready for the next encounter in early July to do more than prepare a three-paragraph description of the discovery image for the imaging team's paper, which was submitted to *Science* 5 weeks after encounter. But 2 weeks after the encounter, Wing-Huen Ip, Fillius's once-reluctant ex-student, submitted his analysis of Fillius's published charged particle data to *Science*'s rival, *Nature*. Among other conclusions, he found that at least some of the ring particles must be meter-size boulders or even larger to have stopped charged particles the way they did. Ten days later, Roman Smoluchowski submitted a paper to *Nature* that suggested that the ring was the debris from the tidal breakup of a weak, carbon-rich body. Bombarded by Jupiter's trapped radiation, atomic-scale erosion was slowly shrinking the smaller particles to the point that radiation forces would remove them, he concluded.

Thirty-eight hours and 1.5 million kilometers from closest approach to Jupiter on 9 July, and just 2.5 degrees above the ring plane, *Voyager 2* took the first 2 pictures of the 24 it would record. These 2 were the first ever to look down on the ring rather than at its edge. There was only a single, faint, rather narrow ring with no visible gaps subdividing it. Twenty-four hours out, 10 pictures at ring plane crossing confirmed the location of the outer edge at 1.8 Jovian radii. After slipping beneath the ring plane and swinging wide of the planet, Voyager was about to cross back through the ring plane when it made a series of images from within the shadow of Jupiter. The first of these solar occultation images were searches for aurora and lightning on Jupier's night side. But what struck those few who happened to be watching the television monitors around supper time the day after closest approach were the pairs of bright ribbons that jutted from either side of Jupiter. The ring that was faint and tenuous when viewed from the Earth-facing side of Jupiter was bold and bright when seen from the far side. The next 7 narrow-angle images, this time intended for the rings, caught them in their full glory.

These views back toward the Sun from within Jupiter's shadow were the real payoff. The perspective was intentional, but the ring's brightness had startled Voyager's operators nonetheless. It was a unique view. Earth-bound astronomers must always observe the outer planets in light reflected back to them by ice, rock, and clouds. Voyager scientists also first saw the ring by such back-scattered light. So they used *Voyager 2* to catch the ring as sunlight filtered through it, adding a dazzle to it much as the headlights of an oncoming car create a blinding brilliance in a fog. The effect depends crucially on the size of the particles and their number; the smaller they are, the brighter they will appear in forward-scattered light. The Jovian ring, which to an approaching astronaut's eye would at best have the faint tenuousness of a comet's tail, is 20 times brighter when viewed from the far side in forward-scattered light.

The dustiness of the ring implied by its brightness in forward-scattered light presented a problem for theorists. At *Voyager 1*'s flyby, Smith had suggested that the ring particles were relatively large—a ring of dust would be swept away by solar radiation and plasma drag in only a few thousand years. In fact, in back lighting it appeared that the 7,000-kilometer-wide bright ring was even now dribbling dust inward through a faint, wispy sheet presumably extending to Jupiter's upper atmosphere. If the ring were not a short-lived, transitory phenomenon—a philosophically distasteful idea to most planetary scientists—what kept it from being wiped out? The ring needed a source for its continuous renewal and rejuvenation, everyone agreed. But where could such a source be found?

As reported in the *New York Times*, Stone and Smith offered two possible sources in the course of the encounter's instant science. One was internal and one external to the rings. Their possible internal source was the debris from the tidal breakup of a satellite-size object. Sweeping forces would not budge these large chunks, but Jupiter's trapped radiation would blast them unmercifully, turning their surface rock to dust. Too small to hold the dust by gravity alone, the boulders would release it to form an enveloping haze that would eventually drift toward the planet as a sort of petrological dandruff. An external source might be Io, the nearest major satellite. Perhaps the volcanic plumes of that astoundingly active satellite were lofting microscopic debris toward the ring, where some unspecified force would cause it to pause before falling into the planet. This was Stone and Smith's version of Owen's *Voyager 1* press conference suggestion of an external source for the ring.

The quick science this time would have to be quick indeed. All of Voyager's best pictures were plastered across everything from *Aviation Week* to *Science News*. Anyone outside the imaging team could spin theories from them as they pleased. Team members' fear of being scooped was not entirely unjustified. Only a month after *Voyager 1* discovered active volcanoes on Io, Guy Consolmagno of the Harvard-Smithsonian Center for Astrophysics submitted a paper to *Science* in which he argued that the driving force behind the towering volcanic plumes must be the element sulfur. His haste and his multiple citations of a *Washington Post* news story irritated some Voyager scientists and incensed others. Consolmango had not pilfered any Voyager data—the NASA public relations machine ensured that the crucial information was public knowledge—but common etiquette seemed to require a decent interval during which team members might recoup their years of Voyager preparations. Rather than waiting for the traditional group submission of Voyager papers to *Science*, Owen, Danielson, Cook, and the technicians assisting them submitted a paper to *Nature* in late August. In it, they presented their preliminary analysis of the ring's structure, which held up well in its broad aspects, and their best thinking about the ring's origin, which did not fare so well. With more time for their analysis, David Jewitt, a graduate student at Caltech, and Danielson later refined the ring's dimensions and added a few discoveries of their own.

The Jovian ring system looks like no other. It is a single ring of several parts that is undivided by any gaps, at least as far as can be seen through the smearing due to Voyager's breakneck speed and the camera's 96-second exposures. The narrow, bright ring has a sharp outer edge at 1.81 Jovian radii. With a brightness 300,000 times less than the planet's, this part of the ring is bright only in a relative sense. Its optical depth is 3,000 times smaller than that of Saturn's C ring, the one as tenuous as the thinnest of terrestrial clouds. The inside edge of the bright ring is broad and diffuse where the ring fades to a wispy sheet five times fainter still that extends to the atmosphere above the cloud tops. By good fortune, *Pioneer 11*, after a billion-kilometer voyage, passed only a few thousand kilometers inside of this inner edge of the bright ring. Jewitt and Danielson also detected a halo of particles extending, as unlikely as it seemed, 10,000 kilometers

a   **b**   **c**

Figure 5.5
Three faces of the Jovian ring. Processing of one image of the ring seen in forward-scattered light shows (a) the bright ring, (b) the interior faint ring, and (c)
the halo extending above and below the plane of the ring. [Reprinted from D. C.
Jewitt and G. E. Danielson, "The Jovian Ring," *Journal of Geophysical Research* 86
(1981), 8691–8697, copyright 1981 by American Geophysical Union]

above the ring system. After all, rings are flat by definition. These
particles must first be electrically charged due to interaction with
various kinds of radiation, Jewitt and Danielson concluded, and then
lifted to those heights by Jupiter's canted magnetic field as it rotates
at an angle to the ring plane and sweeps up and down through it.
Collisions between ring particles, which are daily events for a particle
in Saturn's denser rings, usually enforce flatness on a ring. But in
Jupiter's inner faint ring sheet, a particle suffers a collision perhaps
once a month. Thus the doughnut effect of the halo.

There was general agreement on another point—the particles seen
by forward scattering are tiny, short-lived bits of rock. By comparing
ring brightness at different viewing angles, several groups calculated
a typical particle radius of a few microns, or only about 10 times larger
than the wavelength of visible light. Hardly the household variety of
dust. Spectra obtained from Earth in visible light suggested that the
particles have a rocky composition, similar perhaps to the rock of
nearby Amalthea. And though the details of the calculations varied,
everyone agreed that these particles are no more than 1,000 years
old. Even before they could be dragged into the planet, some particles
are blasted to smithereens by 140,000-kilometer-per-hour micro-
meteoroids or are chipped away atom by atom by radiation.

In their paper, Owen and his group argued for an external source
to explain the apparent antiquity of a dusty ring. The bright ring does
lie inside the classical Roche limit for a liquid body, they noted, but
it lies outside any realistic Roche limit of even the weakest solid body.
Pollack's argument that lingering gas would drag a primordial, rocky

Figure 5.6
The backlit ring and planet. *Voyager 2* obtained this four-image mosaic as it cut through Jupiter's shadow on 10 July. Micron-size particles in the rings and in a high haze layer in the planet's atmosphere scattered light toward the camera, making them far brighter than when viewed on the incoming leg of the encounter. The upper arms of the ring are cut off as they swing behind the planet and into its shadow. The frames centered on the planet were intended solely for a search for lightning and aurorae in the Jovian atmosphere. [Courtesy of NASA]

ring near Jupiter into the planet still seemed persuasive. And they found likely sources outside the classical Roche limit: cometary debris, meteoroids, debris thrown off the inner satellites by meteorite impacts, and volcanic ejecta from Io. In their model, the bright ring is a sort of planetary traffic jam. Particles never stop moving toward the planet, but they are slowed at the ring and become congested. What slows them down was not clear. Resonances of known satellites seemed too weak. Perhaps, the authors suggested, a combination of electrostatic, gravitational, and resonance mechanisms managed to slow them down.

Joseph Burns could not disagree more. He had not started out to produce an alternative view of the Jovian ring, only to give a seminar on it, as he had promised a colleague at Cornell. But in the course of his preparation, he had rediscovered the curious properties of the ring noted by others and drawn some conclusions of his own. He saw quickly that the tiny ring particles must be short-lived, and therefore had to be replenished. Unlike Owen's group, Burns could see no hope of slowing small particles at the bright ring. So he reinvented Smith and Stone's internal source, with a difference. He knew, as Ip had, that there must be something big within the hazy shroud of the bright

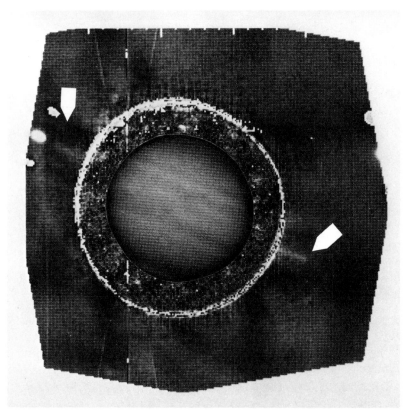

Figure 5.7
The Jovian rings from Earth (ansae—ends of rings—indicated by arrows). This is one of the images of the ring obtained from Earth in the year after its discovery by *Voyager 1*. Jewitt, Danielson, and Terrile used a charged coupled device as the detector on the Hale 5-meter telescope at Mount Palomar. They used a blackened aluminum occultation disk to block some of the glare from Jupiter. Observation through a filter at the methane absorption band also reduced the interference from the planet. At the time (5 March 1980), the ring was tilted 1.2 degrees with respect to the line of sight from Earth. Because the ring signal amounted to only a few percent of the light scattered from the planet, they removed the background glare by subtracting the average of the two strips of adjacent sky from the region of the ring. In the finished image, Jupiter has been inserted for scale within the area of the occultation disk. The ring is the stubby faint line on either side of the disk pointing toward the overexposed image of Amalthea at the extreme left. [Reprinted, by permission, from D. C. Jewitt, G. E. Danielson, and R. J. Terrile, "Ground-Based Observations of the Jovian Ring and Inner Satellites," *Icarus* 48 (1981), 536–539, copyright 1981 by Academic Press]

ring; micron particles could not stop protons with energies of 80 million electron volts fast enough to account for the dips in radiation intensity.

What the ring needed, Burns decided, was a satellite or satellites, maybe 1 to 10 kilometers across, from which small particles could be chipped by 40-kilometer-per-second micrometeorites. The idea was not a new one. Pollack and some colleagues had suggested meteoroid erosion of large particles as a means of reducing initially large Saturnian ring particles to small ones. Burns, though, found inspiration closer to home in the theory of Dermott and Gold, his Cornell colleagues, which asserted that the Uranian rings were small particles shed from a single moonlet embedded in each ring. The small particles could be shattered into the required micron-size grains by the buildup of radiation-induced electrostatic charges. Once formed, Burns reasoned, the particles would be spread to the width of the bright ring by solar radiation forces.

Lo and behold, on 16 October, just a week before Burns was to present his talk at the meeting of the Division of Planetary Sciences on an unseen ring satellite, Jewitt and Danielson announced their discovery in the *Voyager 2* images of Adrastea (1979J1), a moon of 30 to 40 kilometers diameter orbiting near the outer edge of the ring. Jewitt and Danielson considered the possibility that Adrastea supplies particles to the ring, but rejected it because the ring showed no signs of being intensified in the vicinity of the moon. After his initial elation wore off, Burns too dropped the idea of it acting as a source. A single body, he realized, could not explain the considerable width of the bright ring—there must be many small moonlets, even boulders, hidden in the ring. Jewitt and Danielson did see as significant Adrastea's position at the outer edge of the ring. It could well be a shepherd, the first of its kind ever seen. Perhaps Goldreich and Tremaine were right about the Uranian rings. Through its gravitational effect on either the tiny ring particles or the parent bodies that produce them, Adrastea appeared to prevent the outward spread of ring particles and thus maintain its sharp edge. Later, Steve Synnott of JPL discovered Metis (1979J3) orbiting near the outer edge of the ring, another possible shepherd/ring particle source.

Once the idea of an internal source had found an advocate, the planetary science community accepted it readily, if in somewhat altered form. The parent bodies could be anywhere from a centimeter or a meter—depending on who calculates the Poynting-Robertson drag—to a kilometer across, the lower limit of detection by Voyager. Burns, joined by graduate student Mark Showalter, Cuzzi, and Pollack, favor ordinary meteoroids as the eroding impactors. Others have suggested

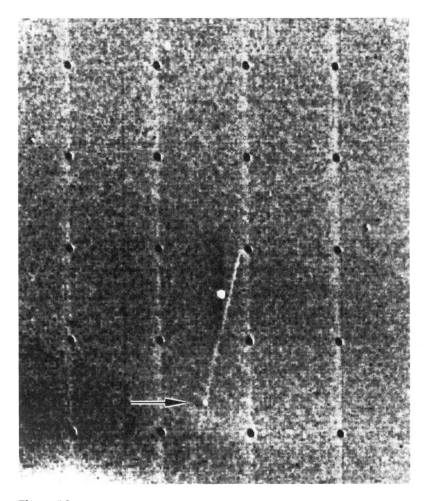

Figure 5.8
A new Jovian satellite—a shepherd? a source? *Voyager 2* caught Adrastea (1979J1) as the spacecraft crossed the ring plane during its approach to the planet. Jewitt noticed the white dot indicated by the arrow and identified it as the image of a satellite orbiting near the outer edge of the ring—the fuzzy line—at 1.80 Jovian radii. Synott later identified Metis (1979J3) in another image orbiting at 1.793 Jovian radii, apparently within the bright component of the ring. [Courtesy of NASA]

microscopic debris spewed out of Io's volcanoes. In either case, Burns's group calculated, the resulting micrometer particles account for only about 600,000 tons of rock, or about as much as in a 10-story-high boulder. If the ring has been around for the entire 4.6-billion-year life of the solar system, 10 trillion tons of rock would have drifted, microgram by microgram, into the planet. That is the mass of one 10-kilometer moon. Despite that drain, there appears to be as much as a million billion tons of rock left in the parent bodies, or "mooms" as Burns is fond of calling them (parent, mother, mom, moon, moom— get it?). A single moon of that mass would be 70 kilometers across.

The nagging question of where the parent moons came from in the first place remains unanswered. No one likes the tidal breakup of a wayward moon—a moon should not have drifted inward to that orbit, should not have broken up if it did, and would have remained in a few big, visible chunks. A small moon stranded there as the early nebula dispersed could have been shattered by a single catastrophic collision with a meteoroid of similar size. Or the parent moons themselves might have formed with Jupiter, as unlikely as that may be according to some theories.

6

Saturn Sports a Narrow Ring

As *Voyager 2* pulled away from Jupiter and its ring in July 1979, *Pioneer 11*, rechristened *Pioneer Saturn*, was plunging toward a September rendezvous with the once unique ringed planet. At Jupiter, it had barely looped inside a ring that no one had warned it about. What lay ahead at Saturn? It appeared that Pioneer had no choice but to pass through one or the other of two recently reported rings. Eventually, a discoverer of each of these new rings would be hailed. Only one of the discoveries was real.

Walter Feibelman was not looking for a ring on that October night in 1966. He was not a ring specialist; he was barely a professional astronomer by most standards. A research engineer in electronic imaging for Westinghouse, he was receiving $1 per hour as a part-time observer on the University of Pittsburgh's 76-centimeter refractor at the Allegheny Observatory. On the night of 27 October, he was searching for faint satellites of Saturn at the most auspicious moment—when Earth crosses the ring plane. Then, the rings are seen edge on and the glare of the rings dims. Feibelman's advantage over those observing at the world's great observatories was that at times he did things "that most people think are crazy." On this occasion, while others exposed their photographic plates for seconds or a minute at most, he was making exposures of 5 to 30 minutes.

On that first night, Feibelman thought that he had found nothing new. A streak extending from the planet's overexposed image must be "poor guiding" of the telescope, as he wrote in his observing notebook. But by 14 November, the night of his next observations, he had decided that there was more to the streak than that. To him, it looked like a faint, outer ring of Saturn that extended at least twice the breadth of the A ring. He continued his ultralong exposures in November and December, but in January the thin, fuzzy line of the

Figure 6.1
One of the best telescopic views of Saturn and its rings. Sixteen images made on
11 March 1974 with the University of Arizona's 155-centimeter Mount Lemmon
Observatory telescope were combined to form this picture. The rings appear
tilted 26.9 degrees, which is nearly the maximum opening seen from Earth. Details of the rings include the Cassini division, a slight brightening near the inner
edge of ring A, and a darker shading near the inner edge of B. *Pioneer Saturn*'s
resolution would be at least several times better than this. [Courtesy of NASA]

ring faded away as the rings tilted open. While the rings were nearly
edge on, all of his long exposures, several a night, showed the outer
ring, he thought.

Rather than make an immediate announcement, Feibelman prepared
a paper for *Nature* that appeared the next May. Knowing that using
the fuzzy plates to make fuzzier prints would convince no one, he
presented more objective, electronic scans across the ring in two of
his plates. His D ring, as he called it, showed up as a blip in the traces.
His "tentative interpretation" of the blips as a ring did not make much
of a splash, he recalls. "By and large the reception was . . . I think
everybody ignored it." When they were not ignoring it, they were
saying that the claim "must be viewed with deep reservation," as one
paper put it.

By 1970, Feibelman's claim of a faint outer ring had been joined
by a report of a new inner ring. On the basis of a single photograph
made at the Pic-du-Midi Observatory in the Pyrenees of France, Pierre
Guérin claimed to have detected a ring between the C ring and the
planet that was only 20 times less bright than the B ring—a real
dazzler compared with the purported outer ring. Guérin also found a
distinct gap between his ring and the C ring. To give his ring a name,

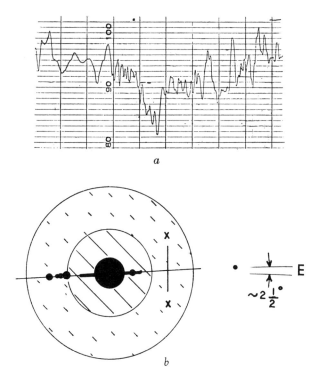

Figure 6.2
The best evidence for Feibelman's outer "D" ring. Part (a) is a tracing of the density of the developed image on a photographic plate along the thin vertical line X———X, as shown in part (b). Density increases downward. Feibelman exposed the plate for 30 minutes on 14 November 1966 when the rings were edge on and, by chance, the bright satellites were bunched on one side of the planet. In (b), the central black circle represents Saturn, the inner cross-hatched circle indicates the greatly overexposed area of the plate, and the outer cross-hatched area represents the faint photographic halo that extends beyond that. Feibelman reported that "a noticeable increase in density at the position of the thin horizontal line is consistently evident on some thirty scans at different positions." The thin line lay exactly in the plane of the rings, that is, inclined about 2.5 degrees to the horizontal. [Reprinted, by permission, from W. A. Feibelman, "Concerning the 'D' Ring of Saturn," *Nature* 214 (1967), 793, copyright 1967 by Macmillan Journals Limited]

he appropriated Feibelman's D designation; the gap became known as the Guérin gap. This D, unlike the outer D, or D′ as it was known for a while, found the academic community reasonably receptive. After all, with a little coaching, anyone could see the gap and ring in his plates. Pollack reproduced Guérin's discovery photograph in his 1975 review of the rings, pointing out the prominent gap in particular. In the whole 90-page paper, Pollack made no reference to Feibelman's report. In a 1971 paper, Franklin and others even suggested a possible satellite resonance to clear the C–D gap, one involving Titan and the gradual reorientation of ring particle orbits.

Although some elevated it to discovery status, not everyone considered Guérin's D ring a proven fact. In 1975 Smith would refer to the D ring as "suspected." In the same year, a Jet Propulsion Laboratory study team headed by S. Dallas produced an internal report entitled "The D-Ring—Fact or Fiction." Working from enlarged copies of Guérin's images, the team studied every aspect of the question, from image processing to resonance calculations. Under "General Conclusions," the team reported that "the best radial distance from Saturn for a spacecraft passage within the visible ring system is through the Guérin division at approximately 72,000 km." That is not to say that the team agreed that the D ring could even be detected. The report is rife with internal contradictions, from the role of resonances in creating ring structure to the limits on the brightness of any D ring. Obviously, not everyone was yet convinced of D's reality.

As early as the 1973 ring workshop, the existence of faint rings outside and inside the traditionally recognized main rings was attracting more than academic interest. A collision of ring and spacecraft could be disastrous. Pioneer would arrive first, but the real concern was for the more highly prized Voyagers, which would pass through Feibelman's D′ ring, if D′ existed at all. At the ring workshop, the reality of D′ received a boost from Kuiper, who had found traces of it in 1966 plates out as far as 6 Saturn radii. However, that was balanced a bit by several other, negative reports. By 1975, Smith, Cook, and Reta Beebe of New Mexico State University had reexamined Feibelman's 14 November plate and concluded "that the evidence of the existence of an outer ring of Saturn remains somewhat shaky." Smith put it a bit more colorfully at the 1978 Saturn system conference— ". . . the outer ring probably still remains, in reality space, somewhere between Farrah Fawcett-Majors and Tinker Bell."

If what Smith's group thought they saw on Feibelman's plate was indeed the Z ring, as they now called the outer ring, its optical depth was about 1/10,000,000. That would make it a million times more tenuous than the C ring, fainter even than the inner faint sheet of the Jovian ring system. Whether such an outer Z ring could wipe out the Voyagers depended on the size of its particles, Smith's group pointed out. Enough millimeter-size particles to produce the ring's brightness would pepper a spacecraft the way a shotgun blast would rip through a ripe melon. But these researchers believed that there would be no particles of that size to greet the Voyagers. Any primordial particles smaller than 7 centimeters would have been removed by the Poynting-Robertson effect, they reasoned, and any meteoroid impact debris swept outward from the A ring by Saturn's plasma could be abundant but micron-size and thus harmless.

While the experts remained somewhat skeptical of Feibelman's ring E, as the game of musical letters finally named it, the inner D ring received convincing support from a bit of modern technology. After unsuccessfully trying to duplicate Guérin's gap and ring in the darkroom through the manipulation of photographic developers and plates, Stephen Larson of the University of Arizona turned a telescope on the D ring that was equipped with a charge coupled device (CCD). More sensitive than a photographic plate and better suited to handling the heavy glare of the planet and main rings, the CCD was also not liable to play the kinds of tricks that photographic developer chemicals can, creating the appearance of sharp, distinct edges where none exist. A methane filter helped cut planet glare, and a mathematical model corrected for the remaining light scattered into the region of D. After all that, an inner ring remained. With an optical depth of 0.02 it seemed to be 200,000 times more dense than the reputed E ring. In contrast with his equivocal feelings about E, Smith could say in 1978 that "today we feel quite confident of the reality of the inner ring. . . ."

Smith's confidence that ring D was real made no difference to the Pioneer scientific team. They wanted *Pioneer Saturn* to blast right into the middle of D, or at least through the Guérin gap between D and C. Smith warned them against it. If the particles were only marble size, 10,000 of them traveling at tens of thousands of kilometers per hour would rip through Pioneer as it tried to cross the ring plane. Even if his best estimates were off by a hundredfold and if Pioneer headed for the gap, it could not survive, Smith warned. To Pioneer investigators, the dangers were secondary. It was their spacecraft, it was a unique opportunity to explore inside the C ring, and besides, Pioneer might just survive, despite all the talk about "ring death" and "kamikaze missions." Pioneer investigators worried that the Voyagers would come along to do more and better science than Pioneer ever could from outside the main rings; an inside trajectory for Pioneer would complement the Voyagers, even if it did not make it through. It was Owen's rationale of moral obligation to explore taken to a lethal extreme.

Despite the scientific team's nearly unanimous preference for the inside option, NASA management in Washington decided that there was too much at stake to exercise this exploration option. If, as seemed most likely, Pioneer splattered itself across the D ring, there would be no first look at Titan, no *Pioneer 11* observations in the outer solar system beyond Saturn, and, perhaps most important, no inkling of what awaited the Voyagers at their ring plane crossings. Both Voyagers would pass through the proposed E ring. *Voyager 2* must cross at 2.86 Saturn radii, just 36,000 kilometers outside of A, in order to receive the gravity assist from Saturn needed to send it on to Uranus. Inside of 3 Saturn radii was unknown ring territory, hidden from Earth-based observers by the glare of the planet and the main rings. The only sure way to probe the Voyager's path and determine just how dense the E ring was there, if it existed at all, was to send Pioneer along *Voyager 2*'s trajectory first. In the end, Thomas Young, NASA director of planetary programs, decided that because it would maximize the benefits to planetary science as a whole, Pioneer would fly outside the rings as a pathfinder for Voyager.

"Close Saturn Encounter May Be End of Pioneer" warned the head-line in the 31 August 1979 *Washington Post*. Datelined Mountain View, California, where Pioneer scientists had gathered at the Ames Research Center for the encounter, the story reported that "scientists said the craft stands no better than a 50-50 chance of surviving the rendezvous." Sounds dramatic, but what John Wolfe, the Pioneer project scientist, had meant was that they knew so little about what lay ahead that anything could happen at the ring plane. The surprises were about to begin, that was for sure. That night, as Pioneer slid faster and faster toward its target point just outside the A ring, it would make the first of its unheralded discoveries.

Midnight was approaching, and Larry Esposito was getting tired. Staring at the television console did not help, with its monotonously regular addition of one image line after another as the spinning space-craft repeatedly swept the eye of its imaging photopolarimeter across the planet and rings. It was an odd view, what with the distortion due to the scanning and the sunshine leaking through the thinner spots from the other side of the rings. The thin spots were easy enough to recognize because of the brightness of the light scattered by what ring particles remained there. But the dark areas could be dense parts of the rings or truly empty gaps, giving the ring images a misleading

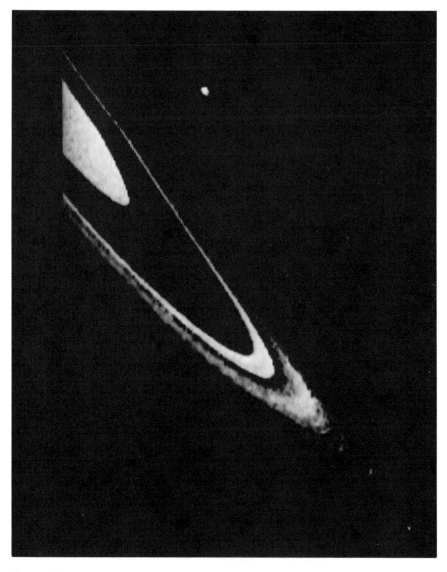

Figure 6.3

Pioneer Saturn's view of the classical rings and the F ring taken on 31 August 1979. This first view of the dark side of the rings reveals the misleading appearance of sunlight leaking through the rings. Regions dense enough to block light, such as the B ring and the inner portion of the A ring, appear black. Thin regions, such as the C ring, parts of the A ring, and the partially filled Cassini division, appear bright due to the scattering of light. Truly empty regions, such as the Pioneer division between A and the newly discovered F ring, are also black. The Pioneer Rock, 1979S1, appears beyond the F ring. [Courtesy of NASA]

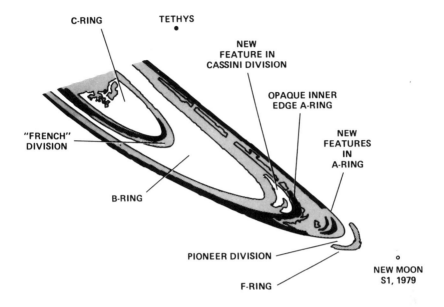

appearance of photographic negatives. Esposito's first interest was photometry of the rings, but in his role as Pioneer's ring man he was at the moment supposed to be keeping an eye out for new rings.

When he first saw it, he was reluctant to call it a ring at all. The more lines that Pioneer added, though, the more the thin, beaded line looked like a ring. Passersby inquired about it, but Esposito declined to push the idea. He later ran into Cuzzi in the hall. "Did you see that?" Cuzzi asked. "It looks like a new ring." "Well, maybe," Esposito conceded. He had to be sure. Rather than wait for machine processing of the data in Arizona, Esposito sat down and began plotting it by hand. By 5 A.M., it was clearly a ring, one so narrow—less than 800 kilometers across—that Pioneer could not resolve it. Centered just 4,000 kilometers outside the A ring, its optical depth of 0.002 or less made it far fainter than A, B, C, or even Guérin's D. The F ring, as the Pioneer imaging team named it, did not seem to be exactly like any other known ring, being wider than a Uranian ring, narrower than the Jovian bright ring, fainter than the Saturnian main rings, and denser than the reported E ring. Still, if it had to have a label, it would be a narrow ring. They were becoming a common breed. Something had to hold this one together too, but what was it?

The next surprise came at midmorning as Pioneer approached ring plane crossing, the first ever at Saturn. Experts took comfort in their calculations of low optical depth and theoretical arguments favoring microscopic particles, but as the moment of truth for the pathfinding spacecraft approached, there was still a fair amount of apprehension about a possible ring death for Pioneer. Donald Humes was manning his console, awaiting any sign of the E ring from his meteoroid detector. Suddenly, things did not look too good for his instrument. Far too soon, the detector recorded its first impact. A minute later, the other data channel of the detector had a hit. Humes was worried—Pioneer was still 900 kilometers above the ring plane. If this was the outer fringe of the E ring, the core of the ring could spell the end of the detector's usefulness; it might not be too healthy for the spacecraft either.

Temporarily silenced by its mandatory postimpact dead time, Humes's detector could give no hint of what awaited Pioneer at the ring plane. Only the continuation of its radio signal would prove that a spacecraft could survive penetration of the ring plane at this distance. The best estimate of the arrival time of the signals sent at the crossing, 10:28 A.M., came and went, but the signal continued. After another minute, just to be sure, Pioneer was declared safe and *Voyager 2*'s route to Uranus clear. Later, Humes could calculate that if this was Feibelman's E ring, it must indeed be tenuous, having a minimum optical depth of about 1/100,000,000, but it was surprisingly thick, presumably 1,800 kilometers thick. That is thick even by Herschel's standards.

Pioneer breezed through the ring plane, but minutes later it reported a sharp, 8-second drop in radiation intensity followed by ripples in the magnetic field. Applying the lessons they had learned in charged particle beam astronomy at Jupiter, project scientists concluded that Pioneer had unknowingly been aimed a few thousand kilometers below and behind a 200-kilometer satellite, the same one the photopolarimeter had detected the night before in the F ring discovery image. Indeed, Pioneer Rock, as it was dubbed, may have been the fabled Janus. It and perhaps a second object were sighted from Earth, but imprecisely positioned, during the 1966 ring plane crossing.

Little was made of it in the press at the time, but this discovery of a small satellite held promise of interesting times for ring specialists. The year before at the Saturn system conference, Giuseppe Colombo had predicted "the existence of a swarm of small bodies between Mimas and the ring A of Saturn." At a distance of 2.5 Saturn radii, the new satellite was well inside Mimas's orbit at 3.1 Saturn radii. What Colombo expected to find was another Mimas—not in terms of size but in the role its resonances seemed to play in shaping the rings. What influence a satellite like Pioneer Rock lacked in mass it might make up for by its proximity to the rings. Several such moonlets would be interesting indeed.

Once Pioneer had safely navigated Saturn's rings, satellites, and radiation belts, the photopolarimeter team had a chance to decide exactly what could and could not be seen in Pioneer's images. With a resolution at least several times greater than any Earth-based observations, it was the best look at the rings ever. And unlike the visual observers of the past few hundred years, these researchers had a permanent, objective record to study. Of the two modern rings, E and D, they could see nothing. E's invisibility could be explained on the basis of its tenuousness, but D's purported optical depth of 0.02 should have made it an easy target. Something a thousand times more tenuous than that might have escaped detection, but Guérin's dense D could not exist.

If that was not a ring set off by a distinct gap on Guérin's photographic plate, what was it? Tom Gehrels and his photopolarimeter group suggested that the "ring" only existed in contrast with the "gap." It, in turn, was probably an adjacency effect, the heightened contrast produced at the edge of a region of higher exposure (the C ring) by the molecular diffusion of chemicals during development of the plate. But Larson's CCD observations involved no photographic plates and no chemicals. What was that D ring? It may have been scattered light, Larson says. His mathematical model was supposed to subtract all the scattered light. But he had assumed a symmetrical pattern of scattering. The spindly supports holding the secondary mirror in the telescope's light path probably made the scattering asymmetric. Thus, the first discovery of the D ring and its confirmation were illusions.

There was no doubt about the discovery of the narrow F ring, the charged particle experiments having detected its radiation shadow each time Pioneer passed beneath the ring. Both imaging and the charged particle detectors suggested an unevenness of brightness in the F ring, as if it had moons embedded in it or the ring particles were clumping together. Electron absorption data even suggested that there were irregularities across the width of the ring.

Imaging also found irregularities in the A and B rings. Ring A had some obvious brightness variations across it, including a particularly dark and thus dense band just outside the Cassini division. Perhaps Earth-bound visual observers had not imagined all of the detail in their drawings. From its unique vantage point, Pioneer could also report that the dark side of the B ring was 10 times brighter than it should be, judging from Earth-based observations of the bright side.

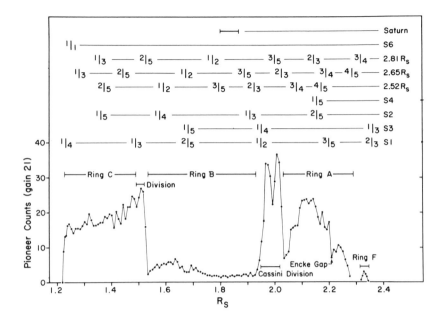

Figure 6.4
Brightness of the rings on the unilluminated side. Although this brightness curve does not resemble those of the more familiar sunlit side, it does reveal new structure and confirm earlier suspicions. Ring A appears to be denser near its inner edge, and the B ring seems a bit thinner near its inner edge. The Cassini division is filled with material that scatters light as well as a denser feature near its center. Both the A and C rings show some small-scale features. From the alignment of satellite resonances, the team concluded, it would appear that Mimas (S1) and Tethys (S4) dominate the broad-scale structure. [Reprinted, by permission, from T. Gehrels et al., "Imaging Photopolarimeter on Pioneer Saturn," *Science* 207 (1980), 434–439, copyright 1980 by American Association for the Advancement of Science]

Light must be leaking through spots where ring particles were less densely packed. The thin spots in B, the densest of all rings, must have an optical thickness less than 0.08 and cover as much as 4 percent of the ring. The rest of B could have an optical thickness greater than 1.5. Although Pioneer could not resolve the light leaks, brightness did seem to vary across the ring, as it would for a series of concentric thin regions.

As for the wealth of structural detail reported over the past 150 years, Pioneer could positively confirm only a tiny bit of it. The Encke gap was there right where Reitsema had reported it on the basis of his Iapetus eclipse observations. There were ring particles within the Cassini division, as a number of observers had suggested. Unimagined by anyone, there was even a banding pattern and a gap in the haze of the division. And the much disputed gap between the B and C rings showed up as a bright band in the shadow of the rings on the planet. Esposito and some of his colleagues concluded that "these new results mean that the ring system is more complicated than expected from earth." Well, more complicated than expected by some. The new complexities "imply that dynamics strongly constrains the location of the ring particles. A placid, homogeneous model of the ring system is not consistent with our data." The explanation for this complexity, Gehrels's group suggested, was the familiar one—Mimas and its resonances, abetted perhaps in some places by Titan.

The Pioneer flyby of Saturn proved to be the heyday of particle beam astronomy. While the photopolarimeter team had laid claim to only one temporary satellite designator, 1979S1 for Pioneer Rock, charged particle experimenters had accumulated five. One (1979S2) turned out to be Pioneer Rock, two (1979S5 and 1979S6) were produced by the clumpy F ring, 1979S3 seemed to be a second satellite at 2.82 Saturn radii, and 1979S4, it was eventually decided, was not an absorption feature after all.

In addition to these observations, one charged particle experimenter reported the discovery of a new type of planetary ring, a planetary asteroid belt really. At the final Pioneer press conference, James Trainor of the Goddard Space Flight Center announced that the absorption of cosmic ray protons detected by his instrument had revealed a new ring, which he called ring G. (As the history of the D-D'-Z-E ring makes plain, names for new rings are there for the taking. Satellites, however, must wait for an international bureaucracy to confer an official designation.) Trainor's broad, diffuse G ring spanned from 10 to 15 Saturn radii and lay between the widely separated orbits of Rhea and Titan. Unlike all other planetary rings, ring G was far outside any

Roche limit. In fact, Colombo's 1978 predictions had also included "a belt of sparse and relatively small objects" between the very same two satellites. He drew an analogy with the asteroid belt that lies between Jupiter and Mars—the powerful gravity of the larger body (Titan in the Saturn case) could have frustrated the accumulation processes that would otherwise have created a third body, leaving a belt or ring of debris in its place. Despite the pleasing consistency of observation and theory, this discovery was not to last. Even before the teams sent their papers for the special issue of *Science* on *Pioneer Saturn*, Trainor and his group decided that a shift in the solar wind made more sense of their proton data than absorption by a broad ring. An asteroidal G ring would not be heard from again, but both the letter and the ring would eventually be resurrected separately.

As 1980 arrived, the bright rings of Saturn began to dim as seen from Earth, giving terrestrial observers their best chance since 1966 to search for faint objects close to the planet. On 6 February, Smith and his group at the University of Arizona sighted Pioneer Rock. On 26 February, Dale Cruikshank of the University of Hawaii detected an object that had the same orbit as Pioneer Rock, usually a sure sign that the same satellite has been sighted again. Not this time. Cruikshank's moon (1980S3) was on the wrong side of Saturn. There really were two tiny moons sharing the same orbit, like riders on opposite sides of a carousel. Later, Earth-based observers found that these two seemed to be moving a bit relative to each other. The discovery of the pair of coorbital satellites confirmed the suspicion of Larson and John Fountain that the 1966 observations perhaps best fit the presence of two or more satellites rather than one.

The 1980 ring plane crossing was also an opportunity for astronomers to bring all of their newly developed gadgets to bear on the detection of the elusive E ring. The CCD built for the training of the Space Telescope camera team was ready, as were various techniques for reducing scattered light and blocking the direct light from satellites, the rings, and the planet. A half-dozen groups around the world strained their instrumentation to catch the faint wisp of light that Feibelman thought he had recorded 14 years before. Everyone saw it, with ease. In fact, Larson was soon detecting the E ring even when it was not edge on. It extends past the orbits of Mimas, Enceladus, Tethys, and Dione to about 8 Saturn radii, or not quite out to Rhea. Curiously, it seemed to have a distinct thickness that increased outward from the main rings. It was also brightest at a distance of 3.7 or 3.9 Saturn radii—there was some disagreement as to the exact location of the maximum. At 3.9 radii, the distance suggested by William Baum and

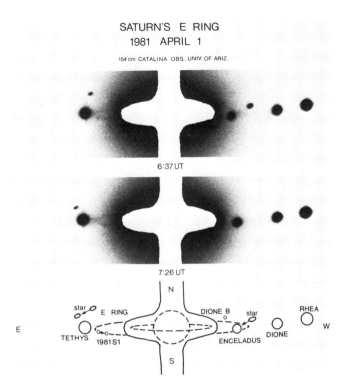

Figure 6.5
A narrow E ring from Earth. This photograph, taken 1 April 1981 by Steve Larson and his colleagues at the University of Arizona using the 154-centimeter telescope of the Catalina Observatory, shows a narrow E ring "core" at 4.10 Saturn
radii. The rings were open 5.4 degrees. [Reprinted, by permission, from J. W.
Fountain, B. A. Smith, and H. J. Reitsema, "Observations of the Saturn E Ring
and a New Satellite," *Icarus* 47 1981, 288–290, copyright (1981) by Academic
Press]

his colleagues, the maximum would be centered on Enceladus. A thick ring centered on a satellite could be a young ring recently produced by that satellite, Baum suggested. If E were more than a million years old or so, he reasoned, it would have collapsed into the ring plane. It apparently had not. And its concentration around the orbit of Enceladus could have resulted from particles being spewed off the satellite, perhaps by volcanoes or a meteorite impact. Theorists had managed to work their way around the need for a young ring around Jupiter. Could they do it here? Perhaps Voyager would help out when it arrived.

There still remained the question of who discovered the E ring. Everyone now conceded that Feibelman had recorded it on his plates, however faintly. He had recognized it for what it was, but the evidence could not convince anyone else. Is that a discovery? It is if discovery entails detection, recognition, and accurate prediction of what others should find. Another would-be discoverer with an apparently convincing case, such as Guérin, may find the prize snatched away from him. The winner is he who is declared right, who always claimed to be right, and who has the objective evidence of his find.

Rings within Rings within . . .

Voyager team members were finding the approaching rings of Saturn to be, in a favorite phrase of scientists, "more complex than had been expected." Expectations had not been high. In 1974, a Voyager working group on the rings expected that Voyager observations would "serve to refine the dimensions [of the rings], especially the location of the features such as the Cassini, Enke [sic], and Lyot divisions. Smaller scaled features of the ring structure have been reported occasionally by earth-based observers and may be further defined by the MJS [Voyager] observations." Other than thoughts of possibly finding some narrow divisions, that was about as far as anyone had speculated in public as to what the rings would look like up close. The prospects for finding new structure in the rings had been improved a bit by recent discoveries—the system of narrow rings around Uranus, the narrow F ring, the light leaking through the B ring, and the small moons orbiting near the outer edge of the A ring. Still, there seemed little reason for idle speculation beyond considering one possibility— that Lyot and Dollfus had not been imagining all of those subdivisions in their drawings of the rings.

By midsummer 1980, it looked as if the Earth-based observers might be right. As Voyager closed within 100 million kilometers of Saturn and the resolution of its camera exceeded the best possible from Earth, the main rings began crystallizing into ever increasing numbers of concentric rings. By October, the complexity of the rings was exceeding the expectations of theorists and observers alike. Cassini had a ringlet and two gaps. Dozens of rings were materializing in all three of the main rings. Had anyone seen these rings before? Perhaps not, but they were reminiscent of Tuttle's nineteenth-century report of "a series of waves in the rings" and Livingston's 1975 description of "numerous concentric internal divisions," more numerous than Lyot's.

Figure 7.1
The rings coming into focus. This image, taken on October 25, shows the new ring structure emerging as *Voyager 1* closes toward Saturn. The Encke division and the Cassini division are obvious, as is the thin material partially filling the latter. Concentric features are numerous in the A ring, the B ring, and especially the dark inner C ring. The two specks just outside the outer edge of the A ring are the F ring shepherd satellites. The black dots are reference markings on the camera. [Courtesy of NASA]

Exceptional viewing conditions on Earth might have given a hint of Voyager's first view after all.

The proliferation of rings was stunning enough, but Voyager uncovered a confounding new phenomenon: shadowy fingers radiating *across* the B ring. According to Maxwell and Kepler, anything in the rings has to behave like a satellite—objects orbiting inside it should overtake it, and those outside it should fall behind. But some of these faint fingers seemed to persist for at least 3 hours without being sheared apart by the difference in orbital velocities along them. For unknown reasons, they seemed to be acting like the rigid spokes of a wheel rather than a swarm of particles. How could these spokes do that? To find out, a sequence of images scheduled for the end of the month was retargeted to focus on the rings.

In the meantime, something would have to be done about all of these new and enticing observations. Consolmagno had got the jump on the analysis of the images of Io's volcanoes, and some of the imaging team members did not want to see that sort of thing happen in the case of the rings. The encounter's publicity machine had not yet been cranked up, giving the team a few extra weeks to prepare a manuscript and submit it to *Nature*. Arriving at *Nature* on 31 October, 3 days after the first press conference, the quickly prepared paper included brief descriptions, a few pictures, and some of the team's first impressions. Of the 32 features that they could list, about a dozen seemed to match the locations of known resonances, including the relatively powerful resonances of the coorbital satellites. Because one of the coorbitals seemed to be catching up with the other, some of the resonance-induced features might transform themselves before *Voyager 2* arrived. Still, many of the new features remained unexplained by the classical resonances.

Perhaps, the imaging team suggested, "small, as yet undiscovered satellites within the rings" could be generating some of the unexplained and unexpected structure. From within the rings, the handicap of small mass would be overcome by the operation of gravity at short range. Even relatively small chunks of ice could create effective classical resonances nearby, or clear gaps in the manner of the Uranian shepherd mechanism of Goldreich and Tremaine. They might even supply new particles to the rings, as Burns had suggested for the Jovian ring. The spokes, in addition, suggested to the team "that other dynamical processes may also be important in producing observed features." This quickest of quick science did get the team on the record concerning some mechanisms vital to the creation of Saturnian rings. Nonetheless, like most quick science, it was seriously flawed—it was too early in

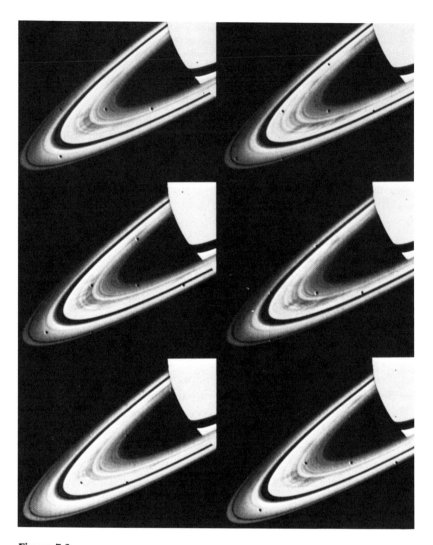

Figure 7.2
Dark spokes on the B ring. A group of dark spokes moves around the B ring in this sequence of images taken on 25 October. The sequence begins in the upper left and moves left to right. Although it was not detectable in this sequence, each part of a spoke moves at a velocity appropriate to a ring particle orbiting at the same distance, which eventually disrupts the spokes. In unenhanced images, the spokes are about 10 percent darker than the surrounding ring. [Courtesy of NASA]

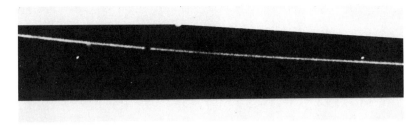

Figure 7.3
Two shepherds and their flock. Satellites 1980S26 (left) and 1980S27 (right) shepherd the F ring on 16 November. The outer edge of the A ring is in the upper right. [Courtesy of NASA]

the data analysis process to have an accurate distance scale for Voyager images, which invalidated all of the resonance-structure correlations.

No sooner had the *Nature* paper been put together than its hypothesis of small satellites shaping the rings received a considerable boost — the narrow F ring had a pair of tiny companions. Like a couple of balls on a giant Monte Carlo wheel, one of these 100-kilometer moons was orbiting just inside and the other just outside the very thin line of the F ring. These must be the ring shepherds that Goldreich and Tremaine could never hope to see from Earth; the F ring truly is a Uranian-style ring. The discovery had been a nice bit of science — observation had led to hypothesis, prediction, and, at least by inference, confirmation. Like the initial observation of Uranian rings, though, this confirmation had a bit of serendipity to it, the shepherds having been discovered in the spoke movie taken on 25 October.

The week before Voyager's 12 November encounter with Saturn yielded new evidence that small satellites (or large ring particles, if you will) can shape the rings. As Richard Terrile, a young astronomer, told *The New Yorker* writer Henry Cooper the day before closest approach, the discoveries were coming thick and fast at JPL:

A few days ago Torrence Johnson — another member of the imaging team — and I were here when a picture of the F ring came down. Because the orbit of its outer shepherding moon, S13, was so eccentric, we thought the F ring itself might be eccentric, too, and when we compared two pictures of it showing its relationship to the outer edge of the A ring, we saw that it was. Then, last night, I saw an eccentric ring in one of the small gaps in the C ring. I noticed it because it was easy to see the ring on one side of Saturn, where it was near the center of the gap, and hard to see on the other, where it had moved over to one side, closer to some other ringlets. We now had two eccentric rings. An hour or so later, I decided to look inside

Figure 7.4
A C ring gap and its eccentric ringlet. A two-image composite picture from images made on 10 November highlights an eccentric ringlet. The images from the two ansae reveal the broadening and radial offset of the ringlet, a characteristic of some Uranian rings. [Courtesy of NASA]

the Cassini division—and, yes, there was an eccentric ring *there*. So it seems that eccentric rings and gaps go together.

As Terrile noted, making discoveries in the age of the spacecraft "is just a matter of who's around." To top it all off, on 8 November Voyager had found another tiny moon just 1,000 kilometers outside the A ring that seemed to be herding that ring and sharpening its edge. Despite a good deal of hype at one of the press conferences about the unexpectedness of eccentric rings, these narrow rings were simply Uranian-style rings.

For real astonishment, one had to go no farther than the F ring. By 8 November, Voyager had confirmed the suspicions of Pioneer charged particle experimenters—the F ring had at least two exceptionally bright sections or clumps as long as 1,000 kilometers that orbited at the same speed as ring particles day after day without changing. Then, at the daily press conference on 12 November, Brad Smith introduced the latest ring images by saying that "in this strange world of Saturn's rings, the bizarre becomes the commonplace, and that is what we saw on the F ring this morning." The image projected on the screen was hardly recognizable as a ring; if it was a ring, it was much the worse for wear. It seemed to be bent, broken, and unraveled. The two brightest strands merged, parted, and merged again in the same image frame, while a diffuse, faint inner ring remained unperturbed inside of the two intertwined rings. The next day, however, the faint component would be seen outside the bright ring.

"It boggles the mind," was Smith's reaction. The kinks and apparent braiding seemed to defy the laws of celestial mechanics, he said. Too late to avoid being widely misquoted, Smith added that "clearly, the ring particles are behaving according to natural laws, it's just that we don't understand [how to apply] them." Gravity appearing to be too unsophisticated a force to explain the weird behavior of the F ring, researchers looked for other forces that might shape rings. Magnetic and electrostatic forces seemed like good bets after Voyager passed Saturn that day and looked back the way it came. In forward-scattered light the F ring was brighter, implying that it contained an abundance of microscopic particles. Those are just the kind, Smith noted, that could become electrically charged and be batted around into odd shapes by Saturn's magnetic field.

Peter Goldreich, for one, had more faith in gravity. "Don't worry about Newton's equations," he said the same day, "most people don't realize how many solutions they have." That afternoon in his Caltech office, Goldreich, with Tremaine and Dermott, explained to a reporter how versatile gravity can be. A gravitational interaction between the two coorbitals, for example, would save them at the last moment from a seemingly inevitable collision. 1980S1, the inner, faster moving of the pair, had seemed as if it would clip 1980S3 when it caught up with it. Instead, according to a theory once applied to the Uranian rings by Dermott, as the pursuer closed in the two would exchange orbital energy and thus exchange orbits. The pursuer would become the pursued. From the point of view of one coorbital, the other would take 4 years to trace a circular, horseshoe path that began and ended in close encounter.

These theorists had nothing quite so neat to explain multiple, interweaving rings, but they saw some reasons to keep their faith in gravity. Collisions were one reason. The classical celestial mechanics of Newton allowed no collisions—the Sun, moons, and planets do not

Figure 7.5
The kinked, clumped, and multiple F ring. These two bright components of the F ring appear to cross each other in this 12 November image, giving rise to their description as "braided." However, there is no evidence that the two components are wound about each other. The bright components also are kinked and contain denser clumps. The sharp break in the ring was caused by a reference marker on the lens of the camera. A faint, inner component appears to be unperturbed by the force that distorts the others. The outer A ring and the Keeler division appear at the upper right. [Courtesy of NASA]

collide in the normal course of events. Collisions between the molecules of water or air, on the other hand, are frequent and essential to any understanding of the behavior of fluids. Ring dynamics falls between collisionless celestial mechanics and fluid dynamics. Relatively frequent collisions are essential to the shepherding of narrow rings, but the damping due to collisions is inversely proportional to the size of the particle. The collisional damping of rocks would take longer than that of dust, a difference that could conceivably segregate them into separate rings. As for the distortions of the F ring components, likely candidates were the shepherds themselves. Their relatively large size might allow them to induce waves in rings. But even Goldreich conceded that the sharp kinks had him stumped.

The F ring was baffling some of the best theoretical minds, boggling others, but it was only one of many rings (some of the time, anyway). Voyager was finding more rings all the time. The *Nature* paper listed 32 features in late October, a half-dozen of which were the classic divisions and boundaries that had been known for 350 years. By 8 November, the official ring count had hit 95. Reporters were discarding the analogy of ripples in the cosmic pond in favor of the more numerous grooves in a phonograph. Decreasing range and increasing resolution were yielding finer and finer details, a common enough principle in spacecraft flybys of planets and satellites. Earth-based and Pioneer observations had hinted that the principle applied to rings as well, but nothing had prepared astronomers for Voyager's revelations. The Cassini division had first sharpened into 4 bright bands, and then those splintered into 20 or more bands and ringlets. As Voyager passed close to Saturn the total "ring" count soared to perhaps 500 to 1,000 — no one had the time or patience to count each and every one.

A thousand rings seemed a monumental problem for theorists. They had run out of resonances long ago. Classical resonances did appear to work in some places — those of the F ring shepherds coincided with what appeared to be dark gaps in A, the Mimas 2 : 1 resonance still fell at the outer edge of B, a Titan resonance fit a gap in C, and there were a few other matches. The successes were a paltry few, though, next to a thousand rings or features, or whatever you called them. There were glaring problem areas, too. The inner edge of C had no resonance, and Encke was empty and sharp edged but had no major resonance either.

If the resonances of the major satellites could not reach inside the rings, the thinking went, perhaps many small "satellites" within the rings could individually shape the part of the ring within their much shorter reach. The preencounter *Nature* paper had suggested that embedded moonlets, as they would come to be called, might shepherd the edges of major divisions; the F ring shepherds had shown that moonlets could "repel" ring particles at short ranges; and minor gaps containing narrow, eccentric ringlets seemed to show that unseen moonlets could orbit within the main rings. Less massive moonlets, perhaps only a kilometer across, might also plow particles aside, only to have collision-induced spreading begin to fill in behind them.

Hundreds of such embedded moonlets might be creating the phonograph texture of bright and not so bright lanes in the B ring, like a regiment of circling snowplows fighting the endless drifting of the snow.

Embedded moonlets looked promising, but even before the ring count leveled off, Cuzzi was cautioning that moonlets were probably not the whole answer. Structure in parts of the Cassini division and the C ring looked too regular to be created by embedded moonlets, and the A ring had a decidedly subdued, more uniform appearance than either B or C. Something more was needed, something Voyager served up as it swooped beneath the rings and peered through the bright, back-lighted Cassini division. As Cuzzi watched this view beamed back by Voyager, a curious regularity in a bright band caught his eye. A pattern of smooth fluctuations filling the band was far more reminiscent of a phonograph record in its regularity than the random patterns of the B ring ever were. Cuzzi thought that he recognized the pattern right away, but in the months following the encounter he, Frank Shu at the University of California at Berkeley, and Shu's graduate student Jack Lissauer would build a case to support Cuzzi's gut feeling.

Shu's expertise was crucial because his specialty was the dynamics of galaxies, and this looked like a galactic phenomenon. Cuzzi had not expected to see any galaxylike spiral density waves because his calculations had suggested that Voyager would not resolve them if they existed at all. But here in the Cassini division the responsible resonance is an unusual resonance of Iapetus, which increased the wavelength enough to catch Cuzzi's eye. No single image could take in the entire, tightly wound wave as it spiraled outward from the resonance like a watch spring, but in the close-up Cuzzi could detect an outward decrease in spacing between bright waves of bunched particles. That was exactly the predicted behavior of a spiral density wave that Goldreich and Tremaine had worked out for rings and suggested might be detectable from spacecraft. The formal announcement of a spiral density wave in the Cassini division would appear in *Nature* the week of *Voyager 2*'s flyby. The team report submitted to *Science* in early February, on the other hand, was a hodgepodge of speculation including talk of possible overlapping classical resonances and converging resonances to explain fine, regular patterns in the A ring. But, in one caption and the closing discussion, it also included the suggestion that certain regularly spaced features in the A ring near the Encke division might be spiral density waves. They might not be so rare after all.

While the ring count was growing at an alarming rate, the simple act of Voyager passing by the planet (executing as it did two distant,

Figure 7.6
The first known spiral density wave. This *Voyager 1* close-up of the unilluminated side of the Cassini division reveals its complex structure of relatively uniform plateaus set off by narrow gaps and wider rifts. The widest and brightest plateau in this view is in the outer Cassini division. (The A ring is to the left.) There is an apparent regularity in its texture that under close analysis turned out to be a spiral density wave propagating outward from a resonance of the moon Iapetus. The tightening of the spiral wave, which resembles a watch spring in form, can be seen in the progressively smaller spacing of the phonographlike grooves with increasing distance from Saturn. [Courtesy of NASA]

utterly uneventful ring plane crossings) garnered several more ring discoveries. The F ring was not the only feature to brighten in forward-scattered light. The E ring revealed itself to Voyager for the first time from the far side of the planet. Ring E must contain fine particles too. Terrile had surmised as much from his Earth-based infrared observations when he and Cook suggested an Enceladus origin for E. As they had anticipated, the startling smoothness of Enceladus's surface — *Voyager 1* could not detect a single crater or mountain at a resolution of 11 kilometers — might be caused by the continual resurfacing of the moon by water from an ice-covered liquid interior. If that were so, a meteorite might punch a hole in the icy crust and expose the liquid water beneath, which would flash to vapor and form tiny ice crystals. That would certainly help explain what a young ring of fine particles was doing all the way out there, apparently in some kind of association with Enceladus.

Forward scattering also led to the second discovery of ring D, the Cheshire-cat-like ring inward of the faint C ring. The Pioneer people were right; this was not Guérin's D ring. "The optical depth is extremely small," reported the imaging team, "and it is very likely, despite reports to the contrary, that the D ring has never been observed from the ground." Whatever holds back the C ring at its inner edge, this vanishingly faint haze of particles seems to leak past it, presumably all the way to the top of the atmosphere. Even here, in this rarefied ring as far from major satellites as a ring can get, there is structure, "numerous narrow features that vary in width from several hundred kilometers to the resolution limit of 3.5 kilometers." Whatever the processes are that shape rings, they are not too particular about what kind of ring they do it to. One feature Voyager failed to find was the gap between the C and the D rings. Pioneer's proposed ring plane target point would have meant disaster considering D's equal brightness in back-scattered light, implying the presence of larger, lethal particles. With the second discovery of the D ring, Saturn's ring system was once again reasonably continuous, if somewhat disorderly, from D on the inside through C, B, A, F, and E.

Voyager had one more ring to add to the system, though. From its vantage point beyond Saturn, it settled once and for all the question of which ring to label G. The asteroidal G between Titan and Rhea had been dropped within weeks of its announced discovery. No sooner was it abandoned than the G label was picked up by James Van Allen of the University of Iowa to designate an apparent ring found by charged particle experiments at 2.51 Saturn radii. The proposed ring turned out to be one of the coorbitals. Now Voyager had by chance

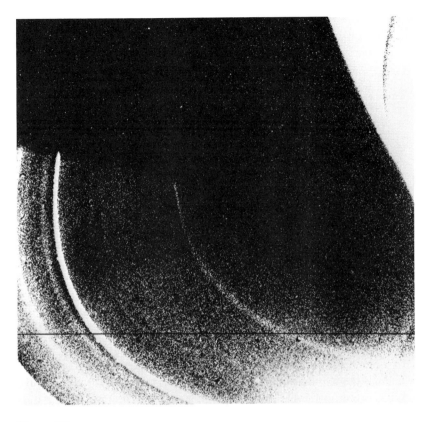

Figure 7.7
The D ring, at last. In this 13 November view, Voyager caught the D ring between Saturn (upper right corner), the inner edge of the C ring (far lower left corner), and Saturn's shadow (upper part of image). Voyager was near closest approach and still "below" the rings on the side opposite the Sun. Thus, D is seen here in forward-scattered light, although it appeared equally bright in back-scattered light. Ringlets are visible in D as they are elsewhere in the rings. [Courtesy of NASA]

caught a faint ring in the corner of the image of the E ring. The smudged image of the ring went unnoticed at first, but it eventually turned out to be at about 2.83 Saturn radii. That was where Pioneer charged particle experimenters had expected to find a moon, 1979S3. The new G ring was also uncomfortably close to *Voyager 2*'s ring plane target point of 2.87 Saturn radii. This G ring was unaccountably narrow, but only about 2,500 kilometers separated its center from the path of *Voyager 2*. Where was the ring's edge, exactly?

Forward scattering contributed one more discovery—what the spokes were, if not what created them. One day they were 10 to 15 percent darker than the rest of the B ring; the next they were brighter than B. That settled it. The spokes were not darker particles or areas of fewer particles; they are microscopic particles that are perhaps elevated above the main part of the ring by some magnetic-electrostatic process. Team members discarded as well the idea that the magnetic field of Saturn somehow carried the spoke particles along at its own rotation rate in defiance of Keplerian laws of orbital motion. Three images shot at close range were enough to reveal the gradual slanting of a wedge-shaped spoke as it began its inevitable disintegration. They were not rigid spokes, but the name would stick anyway. Two observations suggested that the spokes' origin, if not their preservation, was linked to the magnetic field. Spokes always cut straight across the rings or slanted in the direction that Keplerian motion would slant them. And the narrowest part of the spoke's wedge was at the radial distance where the magnetic field and an orbiting particle take the same time to complete one revolution. It appeared that some process associated with the magnetic field imprinted spokes on the B ring but did not do it instantaneously, broadening the spoke where the ring particles moved relative to the magnetic field.

Another apparent clue to the origin of spokes came from Voyager's planetary radio astronomy (PRA) experiment. Intended for study of the radio emissions produced by the interaction of energetic charged particles, plasmas, and magnetic fields, the PRA experiment began picking up an unexpected sort of radio frequency static about 36 hours before closest approach. Its frequency ranged from 20,000 to 40 million hertz (cycles per second), the full range of the receiver, and each burst lasted about 100 milliseconds. It most resembled the static an ordinary radio picks up during a thunderstorm, but Voyager was not cruising among thunderheads in the atmosphere of Saturn. It was on the other side of Saturn's ionosphere, a supposedly impenetrable barrier for the lower of the frequencies being detected by the PRA receiver. James Warwick, head of the PRA team, could see only one reasonable source beyond the ionosphere—the rings. Collisions between ring particles there might build up an electrostatic charge that would be released as a sudden discharge; lightning a million times more powerful than on Earth would streak across the rings, albeit invisibly due to the absence of an atmosphere. The connection between this Saturn electrostatic discharge (SED) and the spokes was strengthened, Warwick noted, by the similarity of the period of SED intensity variations (10 hours and 10 minutes) and the orbital period of particles in the dense, dusty B ring.

Voyager 1's view of the rings was dazzling, but what of the particles that make up these many and varied rings? As expected, they are cold—85°C above absolute zero according to the infrared experiment. The sharp brightening in the forward-scattered light of E, F, G, the spokes, and some parts of the main rings showed that particle size does vary across the rings, at least in the micron range. The earliest preliminary analysis of the radio science occultation experiment also showed a variation in the size of larger particles. Len Tyler's radio science team had observed a classic occultation in which Voyager was the "star" and the 20-watt radio beam it used to communicate with Earth was the "starlight," except that the "star" moved too. By noting how much of the radio signal ring particles blocked and how much they scattered forward, the team found that the "effective size" of particles in the A ring is 10 meters, in the outer Cassini division 8 meters, and in the C ring 12 meters. Those would be real boulders, but the effective size is the typical particle size only if there is a narrow range of sizes. That seemed unlikely. Cuzzi and Pollack had argued strongly for a broad distribution of sizes on the basis of Earth-based observations. In fact, the radio occultation observations also fit Cuzzi and Pollack's power law distribution of many small particles and a few massive ones. The observations fit the predicted distribution, but went beyond it, calling for a significant number of particles much larger than a meter across. Large in size but relatively small in surface area, these boulders could account for much of the mass but little of the reflected light studied so closely by observers back on Earth.

Outside of some assurances about size, *Voyager 1* revealed little new about ring particles. What dirties the ice? Is their shape round or irregular? Are they solid like ice cubes from the freezer or airy like a freshly plowed snowbank? Setbacks with two Voyager experiments frustrated progress on such questions. MIRIS, the more sophisticated version of the infrared experiment called IRIS that flew on *Voyager 1*, had not been ready for either Voyager launch. And *Voyager 1*'s photopolarimeter developed such severe problems en route that it was shut down before encounter. All was not lost, though. By comparing the brightness of the rings as imaged through different colored filters, the imaging team found that the particles of the C ring and the Cassini division, the two most tenuous parts of the classical rings, are darker and less red than those of the A and B rings. No one offered an explanation of what could link ring particles separated by the more than 25,000 kilometers of the B ring.

As *Voyager 1* pulled away from the crescent Saturn and headed out of the solar system, Voyager team members were busy putting together their initial reports for *Science* and preparing for *Voyager 2*'s arrival at Saturn on 26 August. The spectacular ring discoveries of the first Voyager made the preparations particularly hectic. The planned ring observations had to be enlarged and almost completely revised. In the midst of this and their spiral density wave paper, Cuzzi, Lissauer, and Shu had to decide whether *Voyager 2* could detect any of those hypothetical embedded moonlets. If it could, where was the best place to look? The only practical places were where a black background would highlight any moonlets—the Encke division, the gaps in the Cassini division, and the gaps in the C ring. Voyager could not afford to search all three areas. Cassini has the widest gaps of the three, and, conveniently enough, the spoke movie planned for *Voyager 2* would by chance include Cassini. The compromise was made. The only complete search for an embedded moonlet would be in the Cassini division.

The chances of finding one or two moonlets in the Cassini division seemed good. On the basis of Goldreich and Tremaine's theoretical work, Lissauer, Shu, and Cuzzi calculated the size of the moonlets needed to clear the two wider gaps, called rifts, in the inner Cassini division and the three narrow gaps between the rifts. A 29-kilometer moonlet could clear the inner rift that forms the outer edge of the B ring, and one 19 kilometers across could open the rift 2,000 kilometers farther out. Voyager could readily detect these. If the two rift moonlets were held stationary in the calculations, three 8- or 9-kilometer moonlets would clear the narrow gaps. These moonlets would probably be difficult or impossible to detect. Another eight tiny moonlets could be added to account for the wiggles in the optical depth curve of the inner Cassini division. There was one problem. In this arrangement the B ring would overwhelm the inner rift moonlet that formed the boundary with B, pushing this guardian moonlet and its companions outward with the spreading "pressure" of the ring's densely packed particles. The moonlets could not stop until the sparse particles of the nearly transparent Cassini division were compressed to the particle densities of the B ring. Only a ring of equal density could hope to hold B in its place.

The solution, it seemed, was to call on the mighty Mimas to anchor the guardian moonlet through the larger satellite's 2 : 1 resonance, the one traditionally assigned to hold back the B ring directly. Mimas would determine the location of the outer edge of B, as had always been assumed, but the powerful short-range herding of a shepherd would actually hold back the B ring. Together, Mimas, a shepherd in the truest sense of the word, and its diminutive sheep dog would herd the B ring flock. But there was yet another catch. The resonance lock between the guardian and Mimas should increase the eccentricity of Mimas's orbit, but there was no sign that it had. Lissauer and the others wanted to find a way to dispose of the unwanted eccentricity but could not. Still, it was a pretty theory. It provided an appealing explanation for the inner Cassini division and could be extended to explain the structure of B, the gaps and ringlets in C and D, and the Encke gap, within which Voyager had discovered at least two narrow, clumpy ringlets.

Voyager 2 would have the best chance yet to see whether Mimas, an embedded moonlet, or some unforeseen force really makes rings. It would pass closer to the rings than *Voyager 1*, it would be the first spacecraft to view the sunlit side of the rings at closest approach, and the Sun would be shining more directly down on the rings. The next encounter seemed propitious for the imaging team.

More Rings Still

Things looked bleak for the photopolarimeter team. While the imaging team reveled in their spectacular views of the rings and eagerly awaited *Voyager 2*'s encounter with Saturn, the photopolarimeter subsystem (PPS) team was trying to rebuild their battered instrument. The record of the PPSs had been dismal up to now. The PPS on *Voyager 1* had been dead on arrival at Saturn; *Voyager 2*'s encounter with Jupiter had left its instrument in shambles. It was not that they were complicated instruments. A photomultiplier detector sat behind a 15-centimeter telescope with three rotatable wheels inserted between. The wheels overlapped only where the light beam from the telescope passed to the detector, so that each wheel could place one of the openings arrayed around its edge in the light path. When properly aligned, the holes in the aperture wheel would determine the breadth of the field of view, the filters in the second wheel the wavelength to be analyzed, and the polarizer disks in the analyzer wheel the direction of polarization.

Overwork created the first problems for the PPSs. *Voyager 2*'s instrument began its career as a heavily tested spare. At the last minute it replaced the instrument that had begun acting up on the spacecraft at Cape Canaveral. Once en route to Jupiter, the work continued for both instruments as the analyzer wheel stepped from one position to the next every 0.6 second, and the filter wheel stepped every 3 seconds, day after day, week after week, for 3 months and more. After 5 million steps while studying Venus, Mars, and the faint light scattered from interplanetary dust, a thin film of crud built up on the electrical contacts through which the on-board computer sensed the position of each wheel. Once the film became thick enough, the instrument became confused, uncertain of the positions of the wheels or how to reach the intended new positions. In its confusion, it would step wheels erratically until it lost itself. Lab testing of the PPS that was left behind showed team members how to spin the wheels and remove the film, but that was only the beginning.

Next, the PPSs were blinded and then zapped by radiation. Calibration required that each instrument be pointed at a solar reflector mounted on the spacecraft, but an error in the sequencing of computer commands left them staring wide-eyed at it for hours. *Voyager 1*'s was left fixed on Jupiter for hours as well. Photomultipliers can only take so much—the one on *Voyager 1* had 99.95 percent of its useful life burned away, and *Voyager 2*'s lost about 85 percent of its useful life. Then, on encountering Jupiter, radiation blew a transistor in *Voyager 1*'s wheel drive and destroyed some of *Voyager 2*'s computer control. *Voyager 1*'s PPS was useless after that. *Voyager 2*'s made some atmospheric and satellite observations at Jupiter, but Voyager management wondered why any more effort should be expended on a bad instrument.

The PPS team may have had similar thoughts during the dark days from late 1979 to mid-1981 when it was trying to put Humpty-Dumpty back together again. The team and its new leader Lonne Lane of JPL began reprogramming the computer in order to circumvent the malfunctions, the goal as late as mid-1980 being observations of particles in the atmospheres of Saturn and Titan and in the rings. The possible ring studies did include a stellar occultation, an idea proposed for the PPS by Chuck Lillie in 1972. Larry Esposito, now the resident ring expert on the PPS team, at first had no higher expectations for an occultation than Lillie did. It would pinpoint the classic ring features and determine the optical depth of each ring, which had been goals of Earth-bound observers for centuries. That would be about it. But then Pioneer discovered the F ring and light leaking through holes in the B ring. That was encouraging.

Voyager 1 and its thousand "rings" changed everything. An occultation would obviously be worthwhile if the team could deliver a high enough resolution to warrant the appropriation of the observing time of other instruments. All the other remote-sensing instruments were mounted with the PPS on the same scan platform that swiveled them to their targets. Two factors would ultimately determine the spatial resolution of the PPS: the 10-millisecond sampling frequency of the photopolarimeter, which was originally designed to catch the flickers of a star as a planet's atmosphere occulted it, and the spacecraft's speed during the occultation. Eventually, the team convinced everyone that they could achieve the astonishingly high resolution of 100 meters in the measurement of optical depth across all of the main rings, as well as the F ring. Imaging's resolution across the rings would only be 10 kilometers, and even the limited narrow-angle imaging would achieve at best 600-meter resolution. A hundredfold improvement in the resolution of a complete ring scan convinced project scientist Ed Stone to let the PPS follow the entire 2-hour occultation as the star δ Scorpii passed behind the rings in the shadow of the planet.

By accident, Lane found an occultation of even higher resolution to observe. While watching the closing credits of a computer graphics movie of the upcoming encounter, Lane noticed another bright star crossing the F ring and entering the A ring within Saturn's shadow. Much to his excitement, the computer's stars were authentic. And due to the spacecraft's slower speed, this occultation of the star β Tau would resolve ring structure down to 40 meters. Squeezing the new occultation into the encounter sequence took 5 days of night-and-day work in early August 1981, just weeks before closest approach on 26 August.

Voyager 2 encounter observations were looking more promising all the time, but there was, once again, concern about a ring plane crossing. The problem was the G ring. In a ring system a million kilometers across, the G ring had popped up within a few thousand kilometers of *Voyager 2*'s ring plane targeting point. In fact, the G ring's particles, rather than the E ring's, had apparently punctured the cells of *Pioneer Saturn*'s meteoroid detector as it neared the ring plane. Claude Michaux of JPL had the job of estimating what, if any, hazard Voyager might encounter outside the main rings. Both *Pioneer 11* charged particle experiments and *Voyager 1* imaging seemed to locate the G ring at 2.80 Saturnian radii, give or take 1,000 kilometers or so. *Voyager 2* had to pierce the ring plane at 2.87 Saturnian radii in order to get enough gravity boost to reach Uranus. That would put Voyager 4,200 kilometers beyond the center of G. According to Bruce Randall of the University of Iowa, ring particles detectable by charged particle experiments did not extend beyond 2,700 kilometers to either side of G's center. Thus, Voyager would have a 1,500-kilometer clearance, less whatever error there was in the location of the center of G. The clearance, by the numbers anyway, could be as low as 500 kilometers. If the narrow G ring was eccentric, clumpy, multistranded, kinked, or accompanied by minor satellites, as was the nearby narrow F ring, the clearance could be less. That worried some team members. On the bright side, *Pioneer 11* had cut almost 1,000 kilometers inside of Randall's inner edge but suffered no ill effect. To be safe, Michaux recommended that Voyager keep its distance and stay well outside 2.85 Saturnian radii. Despite Pioneer's pathfinding (it had bracketed the G ring with its two crossings but came no closer than 3,400 kilometers to *Voyager 2*'s target point), another spacecraft would be blazing a new trail.

Before it faced the uncertain perils of bodily probing the ring plane, Voyager's cameras had a string of chores to finish. The first was to record the comings and goings of the ring spokes. They appeared sharper than ever, thanks to *Voyager 2*'s 50 percent greater camera sensitivity, a higher sun angle, and more appropriate exposure times based on *Voyager 1* experience. Still, it was far easier to describe the spokes than explain them. A 6,000-kilometer-long spoke could form in less than 5 minutes and fade from view before it had swung halfway around the planet. Spokes formed anywhere around the B ring, between 1.72 Saturnian radii and its outer edge, but they tended to favor forming over parts of the ring that had recently emerged from Saturn's shadow or had carried a now vanished spoke.

Graduate student Carolyn Porco of Caltech and Ed Danielson found another predictable characteristic, a tendency for spoke activity to peak every 640.6 ± 3.5 minutes. That is identical to the 639.4-minute period of Saturn kilometric radiation (SKR), the natural, long-wave radio emission that Saturn's rotation sweeps across space like a beacon. SKR emission, as well as aurora activity, favors a particular sector of Saturn's magnetic field. When that sector points directly toward the Sun, SKR is most intense; when the sector sweeps across the morning ansa, spoke activity there peaks. The mechanism of this linkage between SKR, spokes, and aurora remains obscure. The most popular idea is that the magnetosphere, dust grains in the ring, and the ionosphere form a giant electrical circuit carrying a current driven by plasma as it is dragged across the rings by the magnetic field. Whether that could drive enough current to precipitate spoke formation—another mysterious process—depends on the largely unknown conductivity of the magnetosphere. Perhaps that is where the SKR sector of the magnetic field comes in. It might in some way modify the circuit as it sweeps by the morning ansa, increasing the current flow and the likelihood of spoke formation.

By 3 days before closest approach, something was missing in Voyager's views of the rings—embedded moonlets. Although Voyager would only detect a moonlet as an unresolved point of light in the imaging movie of the Cassini division, 16 hours of repeated imaging should detect moonlets 6 to 10 kilometers across, the exact limit depending on their albedo. And theory said that Voyager should be seeing a moonlet at least that large in each of the two rifts of the inner Cassini division. So Cuzzi and several colleagues scanned the images through the day and into the night. Nothing. They held out hope until the next day, but then it was over; Brad Smith announced that the embedded moonlet search had failed. Its failure was not just a problem for the Cassini division. If 10-kilometer moonlets did not clear these rifts, then theorists had no reason to imagine 1-kilometer boulders scratching out the phonograph structure of the B ring. Once again, the press reported general befuddlement. Peter Goldreich conceded no such thing. The moonlets could be too small to detect and still be massive enough to clear the rifts, according to his calculations. The press still went with doom and gloom.

That same day Smith announced another no-show—F ring "braiding." Close-ups of F showed multiple strands, at least four in all, but no interweaving of the strands. Even when the two shepherds passed each other (called a conjunction), the ring between them refused to unravel. Between the Voyager encounters, gravity had regained the primary role as the shaper of rings—electromagnetic forces had lost their appeal—but conjunctions seemed the most likely time to see gravitational effects. It appeared that the F ring had begun behaving a bit more properly once it knew that it was on camera. Eventually, after the encounter, one frame turned up that had caught a strand splitting and rejoining. That was the only case of interweaving in the 15 percent of the F ring that *Voyager 2* inspected with sufficient resolution, which included where *Voyager 1* had seen it. That makes interweaving less common than thought at first—or *Voyager 2* just did not look in the right places. But the F ring still had a few surprises left.

Smith had no new braids to show, but he did have golden threads within the C ring. Saturn's rings have only the palest yellow tint through a telescope, but three images made with different filters— ultraviolet, blue, and green—can be combined and the result greatly exaggerated to produce an enhanced "color" image. These had revealed subtle color differences not only between the B ring and the bluer Cassini division and C ring, but also across the A, B, and C rings. Three narrow ringlets in the "blue" C ring even had the golden enhanced color of the B ring. That made theorists uneasy, until they learned that the press conference version of the image had not included a consistent adjustment in all three component images for smearing due to spacecraft motion. That explained the golden threads, but the other color variations would not go away. There are also albedo differences. The particles in the A and B rings are more reflective than those in C and the Cassini division, the latter having an albedo of about 0.2 versus an albedo greater than 0.5 for A and B. Cuzzi even

found albedo variations across 100- or 200-kilometer stretches of the
B ring. The composition of particles must vary from ring to ring and
within rings; perhaps some have more of the reddish contaminant
than others. But how did it get this way? Did huge ring-creating
impacts on small moons of differing compositions create the differences?
Could ring forces preserve those differences for billions of years? Or
could differences in particle size or collisional erosion rates create the
differences?

While the press corps was oohing and aahing over strands of gold
in the C ring, the PPS team was discovering a new problem with their
resurrected instrument. Two days before the scheduled stellar occultation, Lane and his team realized that the instrument might not know
how to align the medium-size, 1-degree field-of-view aperture with
the beam of starlight from the telescope. The culmination of almost
2 years of instrument reconstruction could be a case of self-inflicted
blindness. It was not standard procedure, but on the day before the
occultation Voyager received a direct command sequence on how to
spin the aperture wheel, which should remove the interfering crud,
and to line it up with the spot where the star should be. Time would
tell if it would work.

On 25 August, closest approach day, Voyager was falling toward
a point just beyond the A ring deep in Saturn's shadow, sweeping
wide of the planet above the orbits of Rhea and then Dione, finally
gaining a view of where the shadowed rings would occult δ Scorpii.
If the aperture wheel managed to find its way to the proper position,
Lane would see it in the readout at JPL. Esposito was in Boulder,
watching on closed circuit television. When Voyager finally sent word,
Esposito did not have to see the readout. Lane's jumping for joy told
him that Voyager had a clear view of the star.

The occultation was a joy as well. For 2 hours, the chart recorder
spun out a continuous squiggly line that traced the ups and downs of
optical depth as the star flickered across the rings. The PPS cranked
out one measurement every 10 milliseconds, or every 100 meters,
across 82,000 kilometers of ring. Ainslie never dreamed of this when
he glimpsed the flickering of his stellar occultation. Instead of his two
apparent gaps in the A ring or Lyot's dozen or so subdivisions, the
PPS saw 10,000 rings, if such could be called rings. The PPS was
splitting one of imaging's rings into four or five "rings" or features.
The chart trace drew features as small as the 100-meter resolution
of the occultation. No gaps separated most of them, as Cassini separates
the A and B rings.

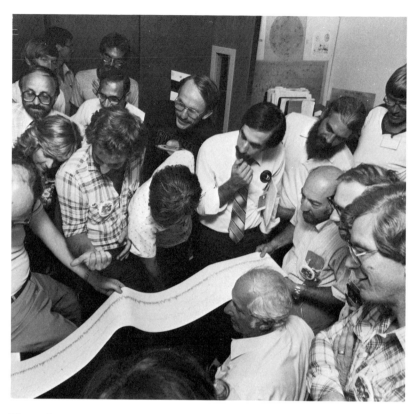

Figure 8.1
Mass viewing of an occultation. What had been private affairs among a few astronomers gathered around a telescope or huddled in the belly of a plane became a public viewing and celebration during *Voyager 2*'s stellar occultation.
Members of other Voyager teams and the press crammed into a tiny room as
team leader Lonne Lane (holding end of chart paper) watched the flickering of
the star represented by the jiggling of the chart recorder's pen. [Courtesy of
NASA]

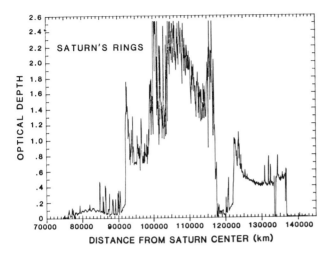

Figure 8.2
Saturn's rings as recorded during the stellar occultation. This is a summary plot at a resolution of 60 kilometers of the 800,000 data points recorded during the experiment, which had a resolution of about 100 meters. It may be compared with Lyot's drawing (figure 2.7) or Westfall's visual occultation observation (figure 2.6), keeping in mind that higher optical depth means a brighter ring and a fainter occulted star. From right to left toward the planet, there are the F ring (the tiny blip just outside 140,000 kilometers), the sharp outer edge of A, the narrow Keeler division near the edge, the Encke division, the higher optical depths near the edges of A, the wide but not empty Cassini division, the dense B ring with its generally banded appearance (note especially the band of lower optical depths near the inner edge), and the C ring and its narrow ringlets. Some optical depths in the B ring exceed the limit of 2.5 of the photopolarimeter. [Reprinted from L. W. Esposito et al., "Voyager Photopolarimeter Stellar Occultation of Saturn's Rings," *Journal of Geophysical Research* (in press), copyright by American Geophysical Union]

The star δ Scorpii did briefly flash through the two true gaps in the outer A ring. One was the thin black line reported by Keeler in 1888, detected by Ainslie in 1917 during an occultation, seen by Lyot in the 1940s, and precisely located by Reitsema's observation of the eclipse of Iapetus in 1978. This is called the Encke division. The second gap was too narrow to have been seen directly by any telescopic observer, but Ainslie seems to have recorded it too during his occultation. Alexander placed Ainslie's detection of it within 500 kilometers of the PPS's outer gap. In a second ironic move, the international nomenclature committee that named Keeler's discovery after Encke named Ainslie's discovery after Keeler.

A point-by-point search of PPS data from 13,000 kilometers of the outer B ring would eventually reveal only one gap there, one paltry

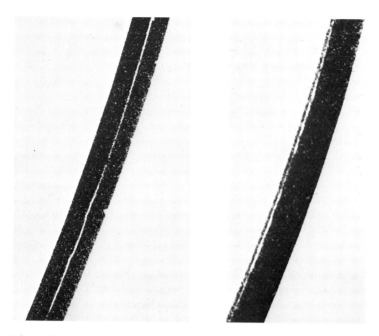

Figure 8.3

Encke ringlets—now you see them, now you don't. *Voyager 2* found that the two ringlets in the A ring's Encke division are incomplete—at some places they are there and at others they are not. In the right-hand image, the inner ringlet is apparent, but the ringlet in the middle of the division seen in the left-hand image is not. Both vary in brightness from place to place, and one may fade from view for stretches as long as 20,000 kilometers, while the other remains clearly visible. Both have kinks tens of kilometers in size. The division is 325 kilometers wide. [Courtesy of NASA]

200-meter empty space. That seemed to rule out embedded moonlets in B; even a 1-kilometer moonlet would supposedly clear a gap its own size. The stellar occultation did immediately help explain a small part of the bewildering jumble of rings coming from the occultation. Among the chaos there was some order—at one point in the B ring, the squiggles became regular and formed a string of ringlets in which each was little narrower and a little more tightly spaced than the one inside of it. The PPS had found another spiral density wave.

Embedded moonlets were down but not out. Voyager's last day this side of Saturn also provided the best resolution that Voyager's camera would achieve. In the Encke division, *Voyager 2* found that the two ringlets discovered by *Voyager 1* are kinky and clumpy—so clumpy in fact that they faded and disappeared altogether for as much as

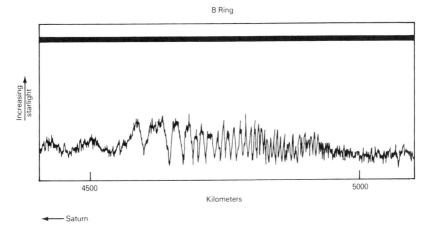

Figure 8.4
A spiral density wave's stellar occultation. This photopolarimeter record of the occultation of δ Scorpii reveals the rhythmic pulsation of starlight created as a spiral density wave in the B ring passed before the star. The shortening of the distance between peaks with increasing distance from the planet is a distinguishing characteristic of a spiral density wave. This one is generated at a resonance of the leading coorbital 1980S1. [Reprinted, by permission, from A. L. Lane, C. W. Hord, R. A. West, L. W. Esposito, D. L. Coffeen, M. Sato, K. E. Simmons, R. B. Pomphrey, and R. B. Morris, "Photopolarimetry from Voyager 2: Preliminary Results on Saturn, Titan and the Rings," *Science* 215 (1982), 537–543, copyright 1982 by American Association for the Advancement of Science]

20,000 kilometers of their orbits. The F ring was not alone in its craziness. Could these ringlets have shepherds too, ones that not only pin the ringlets between them but also hold back the A ring on either side? *Voyager 2* did not search much of Encke, so the required 10- to 20-kilometer moonlets could have escaped detection, if they existed at all. How do you prove the existence of something you have not seen?

As *Voyager 2* began its swing past Saturn and the rings tipped closer to the edge-on view of the ring plane crossing, several of the spacecraft's instruments performed unplanned and most unexpected experiments. The star tracker was not even a scientific instrument, merely a light sensor designed to orient the spacecraft with respect to its reference star Canopus. Although Canopus is one of the brightest stars in the galaxy, something just as bright began distracting the tracker from its intended target—the E ring was getting in the way. Kevin Pang and his colleagues at JPL analyzed the interfering light as well as Earth-based observations and found that it must be reflecting from some oddly uniform particles—they seemed to be tiny spheres of clean water ice of a single, 2- to 2.5-micron size.

Micron-size beads of ice have little future if they are orbiting Saturn. They probably succumb first to micrometeorite impacts and atomic-scale erosion by charged particles. The E ring seen by Voyager must have formed since the Middle Ages, perhaps since the Civil War. But theorists could not explain this young ring as the haze eroded from a core of unseen boulders, as they had in the case of Jupiter. Apparently, a million tons of water had somehow been dispersed as micron droplets and quick-frozen in space.

Enceladus seemed the most likely source, being at or very near the bright center of the ring. Perhaps a meteorite blasted enough water off Enceladus to form the ring. To some, that seemed too lucky a coincidence, an infrequent event occurring at a time convenient for a scientist's theory. The ordinary and everyday would suit most people better. Frequent eruptions of water volcanoes on Enceladus would serve nicely; even *Voyager 2*'s closer look had failed to find meteorite craters on some of the smooth plains of Enceladus, suggesting that some kind of volcanic activity had resurfaced those areas in the geo-logically recent past. The catch is that Enceladus should have frozen solid billions of years ago and stayed that way. Unlike Jupiter's fiery Io, Saturn's Enceladus does not feel the repeated squeezing by the gravity of its parent planet that would melt its interior. Enceladus's orbit is not eccentric enough to permit such repeated tidal distortion. Thus, both satellite geologists and ring specialists had enigmas that seemed to be related. All they needed was a driving force for volcanism, but that was nowhere in sight.

As the star tracker was completing its scan of the E ring and Voyager was approaching the ring plane behind Saturn, another serendipitous experiment began, one that would be called "a limited *in situ* particle detection" in the euphemistic jargon of scientific journals. What hap-pened was that Voyager ripped through the fringe of the G ring. Hundreds of ring particles smashed into the spacecraft every second at 50,000 kilometers per hour. Fortunately for Voyager, these hy-pervelocity projectiles were only 0.6 to 6 microns across, 20 times smaller than the smallest grains that could damage it. Within 10 sec-onds, Voyager was through the 100-kilometer-thick dust swarm and clear of the ring.

No one should ever have known of Voyager's brush with the G ring. The proposal to put a micrometeoroid detector like Pioneer's on board had lost out in the early competition. But investigators for the plasma wave detector were quick to realize that, with no forethought whatsoever, they had conducted a particle detection experiment. The next day at the press conference, Fred Scarf of TRW played back their data recording, whose natural frequencies happened to be in the audible range. The millionfold increase in the instrument's response at ring plane crossing sounded like a hailstorm on a tin roof. Scarf correctly surmised that particles hitting the spacecraft annihilated themselves in clouds of plasma that struck the plasma wave instrument's antenna as they expanded into space. Analysis of the recorded signal provided the first direct, in situ measurements of particle properties since *Pioneer Saturn*'s encounter with two particles 900 kilometers above the plane of the ring.

Even after Voyager's unscheduled probing of its edge, investigators knew little about why the G ring is there. In many ways it resembles the bright part of Jupiter's ring. It is equally faint and narrow. It contains much dust-size, short-lived material and readily absorbs charged particles. But there are no visible moons accompanying it, shepherding it, or resupplying it. It will likely remain a mystery ring for some time to come.

As Voyager shrugged off the G ring's pelting, disaster struck from within. As viewed through the eyes of its cameras, Voyager's world went haywire. Only small chunks of satellites would appear in the corners of images. Instead of taking precious high-resolution images of Enceladus, the cameras pointed into empty space. Finally, the scan platform froze up. It could no longer swing the camera back and forth. The scan platform eventually recovered after clearing its jammed gears, but Voyager could not recover what it had lost. Imaging had missed close-up shots of the F ring, including near-shepherd images and stereo images intended to determine the nature of the "braiding." All of the dark side ring close-ups were lost. And Lonne Lane lost his β Tau occultation of rings F and A at 40-meter resolution. Stone called the mission a 200 percent success despite the losses, and it was— Voyager had seen more rings, in more detail, in more strange configurations than anyone had imagined. Still, the losses hurt the ring people. When would they get back to Saturn?

Two days after closest approach the gloom of the scan platform failure still hung over the daily press conference as Lonne Lane approached the podium to take the next step in a long line of ring measurements. William Herschel had begun by placing an upper limit on ring thickness of 450 kilometers, and the number had been dropping ever since. Observers of Earth's 1966 ring plane crossing reported an upper limit of 2 to 3 kilometers. Some believed that Saturn's rings really were 1 or 2 kilometers thick. Lane squashed that idea. The new upper limit was just 300 meters, which would drop to 200 meters on further analysis. Voyager had not duplicated the traditional thickness determination when the rings appeared edge on at ring plane crossing. There had been too little time. Instead, the PPS team checked the stellar occultation record as δ Scorpii crossed several sharp edges separating dense ring and empty space. Because of the PPS's angled view of the star, the thicker the ring, the longer the transition from obstruction by the full ring thickness to no obstruction. At several sharp edges, the transition took less than 10 milliseconds, requiring an edge thickness less than 200 meters. Tyler and Essam Marouf analyzed the diffraction of the Voyager radio signal at edges and found the same 200-meter upper limit.

If Earth-bound observers were not measuring the true ring thickness, what were they measuring? Even the sophisticated observations of the 1980 ring plane crossing seemed to have detected something that was about 1.5 kilometers thick. In 1979, Joe Burns, Jeff Cuzzi, and some colleagues had warned observers that although they may be measuring some sort of ring thickness, attempts to measure true thickness were futile. As Jeffreys had predicted 60 years earlier, no observer on Earth could ever measure the real ring thickness. It would not take very large irregularities in a vanishingly thin ring to give the appearance of a thick ring, they noted. One irregularity must be the distortion of the rings by the gravity of the Sun and the satellites Mimas and Tethys. They flip up the edges of the A ring hundreds of meters, like the warped brim of a hat. They would warp the E ring too. Its 1,800-kilometer thickness, seen edge on, would add to the apparent thickness of the whole system. And a few embedded moonlets would thicken the rings too.

Voyager 2 found another irregularity in the thickness of the rings—ripples. Unlike the ripples spreading away from a pebble tossed into a pond, these ripples propagate inward, not outward from the disturbance created by the Mimas 5 : 3 resonance. The same 5 : 3 resonance produces the in-plane compression of particles of an outward-propagating spiral density wave, but the slight inclination of Mimas's orbit is enough to drive particles more than 500 meters above and below the ring plane as well. That also had to increase the apparent thickness of the rings. Like spiral density waves, theorists had first sought these bending waves in galaxies.

Bending waves not only helped explain the apparent thickness of the rings as seen from Earth but also provided another means of determining their true thickness. Starting with the heights of the ripples in the Mimas bending wave, Lissauer, Shu, and Cuzzi calculated a

Figure 8.5
One resonance, two waves. On the left is a bending wave, which consists of shadow-casting, 1- to 2-kilometer-high ripples in the ring, and on the right is a spiral density wave. The same 5 : 3 gravitational resonance with the satellite Mimas creates them both, but the gravitational effect of Saturn's oblateness has separated them. [Courtesy of NASA]

true, local thickness for the ring of about 30 meters. In a slightly different approach, Esposito combined the dimensions of the bending wave and an Earth-based determination of the volume density of the ring to calculate a thickness of 30 meters as well. If the depth of the Atlantic Ocean were in the same proportion to its 4,000-kilometer breadth as this thickness is to the breadth of the rings, the abyssal depths of the Atlantic would not quite reach 1 meter; you could never get in over your head.

Peter Goldreich still thought these observers had it wrong — 30 meters was too thick. At the major post-Voyager ring meeting in Toulouse, France, in August 1982, he argued that everyone was looking at the most highly perturbed parts of the rings — bending waves, sharp edges, and spiral density waves. Whatever disturbance creates these features also pumps energy into the rings, he argued, which increases the turbulent motion of particles and puffs up the ring there. The ring would swell to 10 or even 100 times its more typical, unperturbed thickness. Goldreich's best estimate of the true thickness was 5 or 10 meters (for rings that are 270 million meters across).

The middle ground in this debate about ring thickness had actually been staked out before either of the Voyagers reached Saturn. Most recently, Cuzzi, Burns, and their colleagues had argued in two 1979 papers that the expected power law distribution of particle sizes, in which there are a million centimeter-size particles for each meter-size particle, would inevitably produce a thicker ring than Goldreich's. The largest particles, the boulders, may have settled into a monolayer as Jeffreys had argued all the particles had. Still, the relatively powerful gravitational pull of the boulders would stir up the smaller particles during close encounters. Collisions with the boulders would not be required. Once scattered out of the ring plane into a ring many particles thick, the small particles would collide with each other and fall back toward the ring plane, the group concluded. The balance between the gravitational stirring and collisional damping would fix the thickness of the ring at some tens of meters. Most of the mass of the ring would be in a monolayer or a near-monolayer, as Goldreich wanted, but the small particles would form a ring many particles thick, as observation suggested they did. There are two different rings, according to this thinking. One is a core of boulders, which contains most of the mass that dynamicists worry about, and the other is a bright halo of small particles surrounding the core, which accounts for most of the surface area that reflects light back to observers. This view formed an uneasy consensus after Toulouse. No one has shown that it fits all of the data, but it does not seem wildly unreasonable to anyone.

The day after Lane and the PPS team thinned the rings to new extremes, they split the F ring even further than imaging had. The Voyager camera, at a resolution of 15 kilometers, found one bright component in the F ring and at least three or four faint ones. Each was 70 to 100 kilometers across, and together the group was about 700 kilometers across. At a resolution of 1 kilometer, the PPS split the bright core of F into at least 10 individual strands, one of which was itself a dense 3-kilometer-wide core within the core. The PPS's highest resolution revealed optically dense edges to this inner core that reached optical depths typical of the A and B rings. The shape of this occultation trace recalled the similar shape in the occultations of Uranus's η and ϵ rings.

Other instruments saw the F ring a little differently if not more sharply than the PPS. The radio occultation experiment detected only a 2-kilometer-wide strand at a wavelength of 3.6 centimeters and

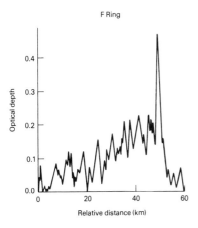

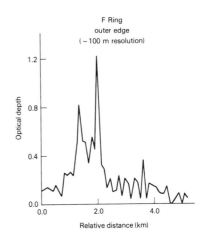

Figure 8.6

A closer look at the F ring. Although the stellar occultation experiment only recorded the changing opacity along a single line across the F ring, its resolution was far greater than that of imaging. A resolution of about 1 kilometer (left), ten times better than imaging's best, reveals profuse structure. The remaining peaks of opacity would appear as strands of fainter material if imaging could have resolved them. On the right, the bright component appears at the maximum 100-meter resolution of the occultation experiment, which splits it into two high-optical-depth components embedded in a dense band. This "W structure" is reminiscent of some Uranian rings. [Courtesy of NASA]

nothing at all at 13 centimeters, prompting Len Tyler to describe F as a core of marbles embedded in a plateau of fine dust. *Pioneer Saturn* charged particle experiments saw a clumpy F ring that was 1,200 kilometers wide, almost twice the width of the F ring seen by Voyager cameras. Perhaps its width varies from time to time. Imaging showed that clumps were spaced roughly every 9,000 kilometers around the ring, which is about the distance ring particles would move with respect to either of their shepherds during one orbit. This spacing lent support to the idea that the shepherds can bunch ring particles into clumps, as did Mark Showalter's computer simulation of a few hundred, non-colliding particles and two shepherds. As the computer-modeled shepherds weave in and out in their eccentric orbits, at times passing within 100 kilometers and less of the ring, they shove particles into clumps, or tighter herds if you will. Imagine the panic in the flock when every 18 years—as predicted by Nicole Boderies, Peter Goldreich, and Scott Tremaine—the inner shepherd actually scrapes the edge of the ring.

Some of the clumps appeared to Terrile to be so dense and compact that he suggested that they were moonlets as large as 10 kilometers embedded in the ring. If true, the F ring contains the full range of particle sizes, from 10-kilometer moons to micron dust. That recalled an idea of Goldreich's. Perhaps the reactions of different-size particles to the inevitable collisions within the F ring could help create some of the ring's odd behavior. Shepherding depends on collisional damping and its erasure of the particles' "memory" of the last passage of a shepherd. But 10-kilometer moonlets have a much longer memory than 1-meter boulders, which in turn have longer memories than dust particles. In time, the largest particles would get out of phase with the smaller ones and "act like a zipper to open up the ring." All particles would still follow Keplerian ellipses; the relation of one ellipse to the next would create the ring's bizarreness. Gravity seemed more inventive all the time.

At the final press conference on 30 August, before a small hard-core crowd of members of the press, Joseph Romig of the PRA team played *Voyager 2*'s version of Saturn electrostatic discharge (SED). This SED was just as intense, but *Voyager 2* heard it less frequently. James Warwick and the other PRA team members still favored a ring source as the best, if not the perfect, explanation. Lonne Lane had even found a single, 200-meter-wide gap in the B ring right at one of the spots appropriate for the mysterious SED generator. Joe Burns had his doubts, though. He wondered what was so wrong with the more mundane explanation of lightning in Saturn's atmosphere; perhaps SED could breach the ionospheric barrier where the shadow of the rings weakened it.

In the end, Burns's suspicions proved justified. A group at the Goddard Space Flight Center headed by Michael Kaiser eventually showed that the comings and goings of SED fit a long, narrow source in the near-equatorial atmosphere, but not a B ring source. They also noticed that Voyager could only detect the low-frequency signals, the ones that the ionosphere was supposed to block, when it had a view of the night side of the planet. The ring shadow might not create a big enough ionospheric hole, but the night hemisphere apparently can. A mystery remains: Why is it that the known Jovian super lightning bolts do not produce JED?

As September arrived, the second and last Voyager was limping away from Saturn, headed for its third ringed planet, Uranus. The astonishment over the proliferation of rings was subsiding as researchers immersed themselves in postencounter analysis and the inevitable preparation of papers, talks, and more papers. *Voyager 2* had recorded the structure of the rings in even more bewildering detail, but, as after *Voyager 1*, no one had seen a ring particle, even as a point of light. A

Figure 8.7
The last look for a long time. As *Voyager 2* pulled away, Saturn presented the
dark side of its rings, the view made familiar by *Pioneer Saturn*. The planet shows
through the tenuous C ring, and the rings cast their shadow across the globe of
the planet. It will be quite awhile before another spacecraft provides the next
close-up view of Saturn's rings. NASA has no firm plans for a Saturn orbiter like
the Galileo mission to Jupiter, which puts the next visit to Saturn at least a decade away. [Courtesy of NASA]

close look at one is not simply a fanciful wish by ring specialists to see the object of all their labors. Now that theorists were considering the effects of collisions, they had to know more about the particles doing the colliding—how large they are, whether they are round or angular, icy hard or snowy soft, resilient or pliable.

Observers cannot tell theorists much about the feel or look of a ring particle, but they finally have a lot to say about particle size. Further analysis of *Voyager 1*'s radio occultation experiment continues the historical trend—the range of particle sizes is greater and extends to larger sizes than anyone had thought. It also revealed a new view of the rings—not only color and optical depth vary across the rings but also particle size. Tyler, Marouf, and the rest of the radio science team determined the distribution of sizes by analyzing their 3 billion bits of data two different ways. Smaller, centimeter-size particles act like a sieve, tending to block the 3.6-centimeter signal and to let the 13-centimeter signal pass. A ratio of the signal strength at the two wavelengths provides a measure of the abundance of particles in the centimeter-size range. Meter-size particles forward scatter the radio signal, which can be used as a measure of the scatterer's size.

At these radio wavelengths there is a different set of rings, or rather a variation on the rings seen at optical wavelengths. The proportion of signal blocked by centimeter-size particles varies from ring to ring, across a ring, and even across a few tens of kilometers. It increases from the Cassini division, which seems to lack this size entirely, to the inner B ring, to the C ring. The A ring data plot has a lopsided W shape of sorts, the proportion of centimer-size particles being low near its inner edge, near zero in its inner third, and increasingly larger all the way to its outer edge. Small particles form two W shapes only 200 kilometers across in C ring ringlets. At the 3.6-centimeter wavelength the F ring is a single ringlet 2 kilometers wide or less, but there are not enough particles larger than a few centimeters for the ring to be detectable at 13 centimeters. This size variability is not limited to particles of a few centimeters. The fine powder that forward scatters light is abundant in the dense B ring, and is more common in the outer A ring, in gap ringlets, and near some resonances. From centimeter size to about 2 meters the proportions of different sizes also vary from place to place, requiring a slightly different power law to describe the rate at which abundance decreases with increasing size.

In larger-size ranges, well beyond person-size particles, the power law breaks down. In the 4- to 8-meter, house-size range, there are more particles in some parts of the rings than an extrapolation of the power law from smaller-size ranges would predict. The numbers of

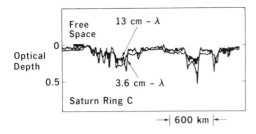

Figure 8.8
Particle size distribution from the radio occultation. The difference in the transmission of microwaves of 13-centimeter and 3.6-centimeter wavelengths (λ) reveals ringlet structure in the C ring created primarily by smaller particles of a few centimeters in size. The double-peaked, W-shaped structure in the opacity curve of the shorter wavelength does not appear in the curve of the longer wavelength, suggesting that a predominance of centimeter-size particles is responsible. [Courtesy of NASA]

these large particles also increase outward across the rings, from about 6 in a square 100 meters by 100 meters in the C ring to 65 in the same area in the A ring. They are relatively scarce, but still the mass of the rings is concentrated in this size range. It is there and not in even larger particles because above about 10 meters there is a sharp cutoff in the size distribution. Particles of 10 to 30 meters, ones the size of a small office building, are missing. Thus there was no evidence of ring particles larger than 10 meters—the radio occultation detected none between 10 to 30 meters, the PPS occultation saw only one gap in B larger than 100 meters, and imaging saw nothing in the Cassini division rifts larger than about 5 kilometers. The coverage is spotty, but huge particles and small moonlets, the kind of objects that would be left behind after a ring-forming, catastrophic meteorite impact, seem hard to find.

No one has seen a ring particle, but that has not stopped some scientists from sketching a picture of what they might be like. Perhaps the most detailed close-up of ring particles comes from a group headed by Stuart Weidenschilling at the Planetary Science Institute in Tucson. They start with a conventional enough view. Say a marble-size particle, whizzing around its orbit at 50,000 kilometers per hour, encounters a 3-meter boulder of a particle nearly square in its path only meters away. Collision is inevitable, but it will be a while coming. At the snail's pace of 4 meters per hour, the marble approaches the boulder and begins to feel the pull of its feeble gravity.

When the marble finally nudges the boulder at the breathtaking speed of more than 5 meters per hour, the marble stays where it hits.

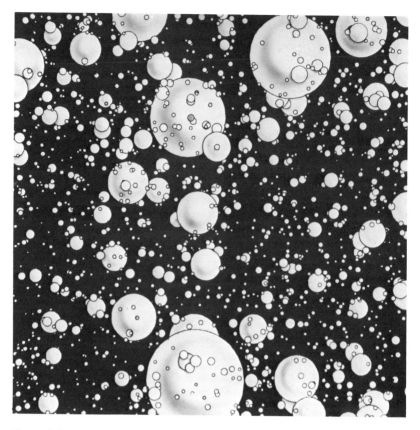

Figure 8.9
A computer's close-up of the A and C rings. A computer program converted size information from the radio occultation experiment into a representation of a square 3 meters by 3 meters of the A (p. 166) and C (p. 167) rings. The particles range in size from 2-centimeter marbles to 70-centimeter beach balls. Larger particles are too rare to have been caught in these squares, there being only one particle in the 4- to 8-meter range in a square 10 meters by 10 meters of the A ring. A closer look at the A ring would reveal many particles too small to be shown here, but the C ring seems to lack smaller particles. [Courtesy of Jet Propulsion Laboratory]

A real marble would bounce off a real rock boulder almost as fast as it hit, but ring particles are probably not hard like marbles and boulders. The Tucson group believes that the surfaces of ring particles, especially the larger ones, are more like that of a snowball, soft and cushiony, than that of an ice cube. Instead of ricocheting around like billiard balls in a pool break, such ring particles would cling to one another whenever they collide. They need not stick to each other—a large particle can hold a smaller one through gravity alone. The smaller particle hits with enough energy to escape the gravitational pull of the larger one, except that the icy rubble on the surface soaks up that kinetic energy during the impact and prevents the small particle's escape. In their model, a small particle could not survive much more than a week before a 3- to 5-meter boulder gobbles it up.

Day by day, a medium-size boulder would grow at the expense of marble-, baseball-, and basketball-size particles, doubling its mass in 10 days. Instead of an unchanging, hard sphere of ice, this sort of particle is a three-dimensional collage of bits of all sizes, a more or less round but lumpy assemblage that would fall into thousands of pieces with a good shove. Depending on the ring, spinning more than one time around in a day or so could slough off parts of its loose "soil" of small particles. Larger particles could be enveloped by a cloud of particles, some escaping, some ever so gently falling back, some making slow-motion bounces across the globe—shades of Jeffreys's rolling avalanche of ice in orbit.

The growth of such ravenous particles could not go on forever; if it did, there would be no range of particle sizes, only moonlets, and besides, Saturn's gravity must eventually tear growing moonlets apart. The Tucson group suggests that the 10-meter cutoff is the size at which this tidal disruption becomes particularly effective. To preserve the observed particle size distribution, a sundered particle must yield the same numbers and sizes of particles that it swept up; the process must be continual but unending. This view of ring particles is not without its problems, but the point is well taken. Just as ring structure is incredibly more complex than assumed only a few years ago, the picture of smooth, baseball-size spheres of solid ice jostling each other from eon unto eon must be too simple.

9

What Makes Rings?

The Voyager spacecraft produced discoveries and mysteries in abundance. Explanations were slower in coming. As team members and theorists who watched the discoveries pour in from Voyager had a chance to catch their breath, the rings of Saturn began to make a bit more sense than the blaring headlines could muster. But even as scientists were eking out some understanding of the appearance of the rings, the recalcitrant mysteries of the rings would bring them back to the age-old question—why are there any rings at all?

The A ring was the most reassuring of the main rings of Saturn. Resonances with the nearby F ring shepherds and the coorbitals seem to create most of the A ring's subdued structure, according to the PPS record of the stellar occultation. While spiral density waves whirl across much of the A ring, few of the stronger near-ring satellite resonances reach as far as the B ring. Because resonance forces drop precipitously with increasing distance, these moons cannot extend their influence much inside the Cassini division. The PPS team managed to identify about 60 spiral density waves in A, each of whose crests might have crossed the occultation path up to a dozen times. Still, the team expects that there cannot be more than 100 waves in all, or at most 1,000 features across the A ring. That is 1,000 out of the 10,000 features across all the rings.

The Cassini division, where some Earth-based observers might have expected to see nothing at all, presented far greater problems than the A ring did. Cassini's gaps, ringlets, and broad plateaus had fit the embedded moonlet hypothesis so nicely. If embedded moonlets had worked there, everyone would have felt better about the A ring's Encke and Keeler divisions and even some of the narrow features of the C and D rings. But *Voyager 2* spoiled that approach, or so it seemed immediately after the encounter. The week before the Toulouse meeting, Esposito and his colleagues submitted a paper to the journal *Icarus* in which they presented a clue to what keeps at least one gap open. They had been matching spiral density waves and the resonances driving them, but they could not link one wave, located just inside the Encke division at 2.22 Saturn radii, with the resonance of any known moon. Perhaps an unseen moonlet within the Encke division was driving it, they suggested.

Then, at Toulouse, Cuzzi and Terrile announced their independent discoveries of Encke's wavy edges. The scalloping's 1,500-kilometer spacing and 5- to 10-kilometer amplitude required one or more moonlets within Encke. In Cuzzi's three-moonlet arrangement, the moonlets would clear three gaps and pinch the two narrow, discontinuous ringlets between them. Angular momentum could move outward unimpeded, from the inner A ring to the inner moonlet, the inner moonlet to the inner ringlet, and so on to the outer A ring. No one had seen scalloping elsewhere, but Tremaine cautioned that other gap moonlets need not be so massive and thus might not noticeably ripple their gap edges.

Moonlets began to look like a good bet for Encke and, by inference, other ringlet-filled gaps. A close look at the ringlets themselves further strengthened the case for embedded moonlets. An 11-member consortium organized by Esposito found that the C ring ringlet at 1.45 Saturn radii in the Maxwell gap bears a striking resemblance to the ϵ ring of Uranus. It is eccentric, one focus of its elliptical shape is at the center of the planet's mass, it precesses without disrupting itself, its width varies from about 30 to 100 kilometers, its edges are less than a kilometer wide, and in cross section its opacity varies in the familiar double peak of the W signature. A working group headed by Porco likewise found a Uranian resemblance in eccentric ringlets at 1.29, 1.95, and 2.27 Saturn radii. The evidence is circumstantial, but ringlets, gaps, and moonlets seem to go together.

Even if the moonlet-gap connection holds up, it still leaves most of the 10,000 new ring features unexplained. How many more might embedded moonlets explain? During the encounter Goldreich had suggested that embedded moonlets might create much of the B ring's chaotic texture. He held to that possibility despite the failure to find gaps of any kind in B. No one understands the shepherding process well enough to exclude the possibility, he said. It is even unclear to him that small moonlets must open an empty gap. In addition, embedded moonlets might shove aside larger particles, which respond mostly to gravitational forces, while smaller particles, whose larger ratios of surface area to mass make them more responsive to the effects of collisions, could diffuse back into the cleared path. Such size-dependent behavior seems to explain the bunching of smaller particles near ringlet edges that creates W signatures. In the Cassini division, moonlets would have to be only a bit more efficient at clearing gaps than calculations show to fall below Voyager's limit of detection. The best current theorizing might have enough holes in it to let embedded moonlets slip through, according to Goldreich.

Perhaps B ring embedded moonlets could not be eliminated, but a new theory has usurped their position as the current speculative favorite. The newly favored mechanism is diffusive instability, as proposed both by Douglas Lin and Peter Bodenheimer of Lick Observatory and by Alan Harris and William Ward of JPL. They suggested that ring particles might spontaneously collapse from a homogeneous sheet into ring features like those in B. The collapse might start with a small disturbance of the ring particles that compresses some of them outward. All the particles continue to transfer angular momentum outward through Keplerian shear, the inevitable collisions between particles in adjacent orbits, but the transfer is a bit more efficient in the area thinned out by the disturbance than where its particles are bunched. Angular momentum thus meets a bottleneck as it moves from the thinned to the more crowded zone and is deposited in the particles between the two zones. Having more angular momentum means those particles must move outward, bunching the particles there even more. As the process feeds on itself, the once homogeneous sheet becomes unstable and collapses into individual ringlets.

Theorists who thought that the diffusive instability mechanism would work nicely in the B ring received an unpleasant surprise at the Saturn system meeting held in Tucson in the spring of 1982. They had assumed that the theory fit the observations; observationalists relieved them of that misconception before the theorists had a chance to make their formal presentation. Theory predicted dense rings, which was fine, and empty gaps, which was not. By the time of the Toulouse meeting that August, Ward and Harris had a new improved version of diffusive instability. Instead of a single particle size, the rings in this version had two sizes, the smaller particles moving faster than the larger ones and tending to fill in the gaps. It was still far from perfect, but it looked encouraging. "It's a physically correct mechanism," said Goldreich, "and it smells like some form of it might be right for the B ring." A stumbling block may be ascertaining the crucial role of the resiliency of particles during collisions, a property that no Voyager experiment could determine.

Backing off to look at the big picture, Esposito and his University of Colorado colleagues tried to make some sense of the structure of

Saturn's main rings by mathematically searching the Voyager occultation record for periodicity in the spacing of ring features. They could detect the influence of spiral density waves on structure in the size range of 15 to 50 kilometers, especially in the outer A ring. But, overall, ring features prefer no particular scale within any of the classical rings. No one mechanism that produces features of a given size or size range creates the bulk of the known structure. Perhaps a properly refined diffusive instability mechanism could generate the observed range and frequency of occurrence of features of various sizes. Or, they suggested, a combination of many mechanisms—some known and some as yet unrecognized—might suffice. So much for making quick sense of the new features in the rings of Saturn.

Theorists are not even certain why one resonance produces a spiral density wave, a second a sharp edge, and a third one of the more subtle narrow bands or gaps seen at resonance locations. "That is a delicate question," as Tremaine put it. The strength of the resonance, which is proportional to the satellite's mass and drops rapidly with its distance, seems to be an important, but not a sole, determinant. The larger of the coorbitals forms the sharp outer edge of the A ring through its 7 : 6 resonance, but its nearly equivalent 6 : 5 resonance produces only a minor feature.

As fascinating as the fine details of the Voyagers' discoveries may be, Esposito's analysis pointed up an obvious but often neglected observation—the most important segregation of ring particles into features of differing optical depth is the division into the three classical rings: A, B, and C. Most spiral density waves and other such details merely ornament the edifice of the main rings.

So where did the main rings come from? For that matter, why are they or any of the rings of other planets still here? No one can be sure of the answers, and even to begin to grapple with these questions one must go back to the beginning. The solar system began 4.6 billion years ago, and there is a philosophical inclination to place the creation of rings at the same time. In fact, those speculating on the genesis of rings often draw parallels with the origin of the solar system, as did Laplace 150 years ago. If rings, or at least the bodies that eventually formed rings, took shape at the time the planets and satellites did, then the same influences should have shaped them all, the reasoning goes.

The genesis story thought to make the most sense for the solar system begins when a cloud of interstellar gas and dust collapsed into a thick, spinning disk warmed by energy derived from its own gravitational collapse. As contraction continued, the Sun coalesced at the center of the disk and further warmed the nebula, especially within its central regions. Eventually, the nebula began to cool enough to condense the least volatile compounds, such as those of iron, nickel, and silicon. Particles formed, much as raindrops do from cooling moist air, and began to fall to the midplane of the solar nebula disk, growing in size as they fell. Once at the equatorial plane of this solar-system-size ring, gravitational instabilities prompted the coalescence of these planetesimals into bodies large enough to grow by gravitational accretion. Through collisions, which dissipated their energy of relative motion and allowed gravity to bind them, they grew into planet-size bodies. Gas not bound by a planet was perhaps blown away by radiation pressure and a strong solar wind from the young Sun.

If planetary rings are primordial, they formed in the inner part of planetary nebulas much like the solar nebula. Just as the Sun's heat prevented the inclusion of much volatile material in the four rocky planets of the inner solar system, Jupiter's heat of contraction prevented the formation of icy satellites or rings in its inner satellite system. The innermost two of Jupiter's four largest moons have mean densities close to those of rock; its ring appears to be rock, too, in the form of fine particles. Being less massive, Saturn released 10 times less heat than Jupiter. That allowed water to condense near Saturn to form its icy inner satellites and rings. So far so good, but now the problems begin for the cosmogonic approach to ring origins. Another force acts to select material for rings—the drag of a nebula's gas. In fact, when rocky grains were condensing and the planets still contracting, the zone of eventual ring formation was still engulfed by the planets themselves, according to some models. Thus Jupiter's having any rings at all surprised theorists, who had convinced themselves that they could understand why Jupiter should not have rings. And Uranus's darker-than-soot ring particles stand in stark contrast with the ice expected on the basis of its nearby icy moons. If these rings formed concurrently with Uranus, their particles must have changed somehow. Perhaps they only gained a surface coating of dark material, or the volatile components of the planetesimals sublimated away, leaving the darker, less volatile, material behind.

In 1848 Roche proposed another way to form rings that might provide ring particles whose composition would seem out of place. Instead of the planet's gravity preventing nearby material from accreting into moon-size bodies, he suggested, a large body—a passing meteoroid or one of the planet's own moons—might pass close enough to the planet to fall prey to tidal disruption by the planet's gravity. Even in its modern form, in which the internal strength of the intruding

Figure 9.1
Mystery of the B ring phonograph. *Voyager 2* recorded this image of a 6,000-kilometer swath of the illuminated side of the B ring. The main structural features are several hundred kilometers wide, and the fine-scale substructure is about 15 kilometers wide. Although some of the fine-scale structure may be waves, the cause of the bulk of the B ring structure remains unknown. Given sufficient numbers and the proper physical properties, ring particles may spontaneously collapse into such a structure. A theory of diffusive instability to explain this structure in these terms is still under development. [Courtesy of NASA]

body is taken into account, the tidal disruption theory has its problems. For one, a passing meteoroid, even if disrupted, would continue on its way with little of it captured in permanent orbit. The debris from a moon in orbit around the planet would remain in a ring of sorts, but initially its particles would be many tens of kilometers across. It is not clear whether collisions could then reduce these large chunks to small particles. Even Uranus's hypothesized shepherd satellites need be only a few kilometers across, so the origin of its dark rings remains puzzling.

An equally catastrophic version of ring genesis has been championed recently by Eugene Shoemaker of the US Geological Survey in Flagstaff. Instead of tidal disruption, he calls upon annihilation by collision. The inner Saturn system certainly seems to have had a violent history. Meteorite impact craters cover most moons. The innermost of the major satellites, 400-kilometer Mimas, was nearly shattered by an impact that left a 130-kilometer-wide crater. A catastrophic impact may have created the coorbital satellites by splitting a single moon. As Shoemaker points out, inside the orbits of Mimas and the coorbitals, where any satellites would have been smaller, Saturn's gravity would have accelerated incoming projectiles to even higher speeds. And there were probably plenty of projectiles in the early days of the solar system before all the planetesimals accreted or were dispersed.

Theories of origins are always fun to think about, but scientists tend to prefer the doable, the problem open to some line of attack, the phenomenon accessible to understanding through detailed study. One such problem is, What has shaped rings since their creation? If, one way or another, ring material came to be in relatively small particles near a planet eons ago, why do we still see such distinctively sculptured rings rather than broad, bland disks winnowed by the ravages of the near-planet environment? So many forces disperse and destroy ring particles. Collisions between particles in adjacent orbits spread a ring inward and outward. Inward, destruction awaits them in the planet's atmosphere. Even within a ring, meteoroids accelerated by the planet's gravity smash into ring particles, splashing off a thousand times the impactor's own mass in microscopic particles. Radiation trapped in the planet's magnetic field wreaks similar destruction on an atomic scale.

Newly generated small particles should disperse or fall into the planet even faster than their larger parent particles. Solely under the influence of the Poynting-Robertson effect from sunlight, particles having the density of ice, smaller than 1 centimeter in radius, and initially orbiting Uranus at its Roche limit, would spiral into the at-

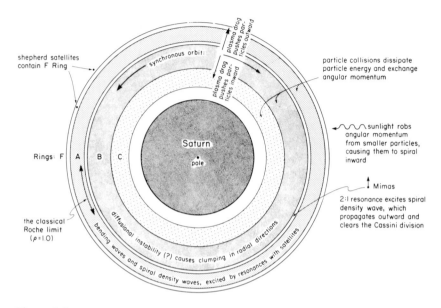

Figure 9.2
The physics of rings. This schematic drawing of Saturn's ring system summarizes the physical processes that influence the motion of ring particles.

mosphere during the 4.6 billion years since the origin of the solar system. In the same time, assuming no other influences, particles smaller than 4 centimeters would spiral inward from Saturn's rings, those smaller than 15 centimeters from Jupiter's rings, and any smaller than 400 centimeters from any ring around Earth. In addition, the drag of a planet's plasma of charged particles would sweep small particles either inward or outward, depending on whether they are orbiting faster or slower than the rotation of the magnetic field and its trapped plasma.

Another vexing problem of ring particle spreading became obvious after *Voyager 2*'s passage by Saturn. Here were these near-ring moons raising spiral density waves in the A ring and in the process robbing ring particles of angular momentum; yet the moons still linger near the ring. The ring particles' loss of angular momentum has not collapsed the A ring inward, and the moons' gains have not propelled them away from the ring, at least not at the rate theorists would predict.

Dynamicists have managed some answers to the question of how ring systems are held together, but only partial answers so far. The ultimate answers should come from a better understanding of the resonance interactions of major satellites and ring particles. As a building requires a firm foundation and a strong framework on which to fasten its weaker components of roof, walls, and floors, so must the delicate latticework of a ring system like Saturn's be based on a sound foundation and a sturdy structure that can withstand erosion processes. The foundation of a stable ring system is the planet itself and its more massive outer satellites. Being far from the planet, the orbits of these satellites change only slowly over the eons. The structure linking the rings and this foundation must be satellite resonances, but a single resonance link would not reach far enough. A chain of resonances is needed.

An obvious first link is the Mimas 2 : 1 resonance that apparently holds back the B ring at its outer edge. Across a distance of less than 1.3 kilometers, one of the densest, most opaque parts of Saturn's rings falls away to nearly empty space. Resonance theory of the early 1970s had a hard enough time creating a gap of any significant width, much less such a sharp boundary. But Nicole Borderies of Caltech, Goldreich, and Tremaine found in the principles of spiral density waves a way to create a barrier to ring particle spreading that is linked to Mimas, 66,000 kilometers away. Instead of Mimas gradually reducing the outward flow of angular momentum, near the resonance angular momentum can flow inward or outward depending on the location along the edge. The net flow decreases to zero, the requirement for an edge, but flows are still high up to the edge, the requirement for a dense ring. The resulting barrier could be so efficient as to have dammed the outward drifting B ring particles for the age of the solar system. In a similar fashion, the 7 : 6 resonance of the larger coorbital maintains the outer edge of the A ring. Regrettably, the other three major ring boundaries—the inner A edge, the B-C boundary, and the inner C edge—have no apparent associations with similar resonances.

Problems multiplied when theorists looked for the next links in the structure. A 200-kilometer coorbital or even 400-kilometer Mimas is far from massive enough to hold back the rings without being pushed outward themselves. Mimas is linked to Tethys, a suitably massive anchor for the rings, through a 2 : 1 resonance. At Toulouse, the experts reported that they had looked for resonance relations between the near-ring moons and Mimas. They found none. They did not even have promising possibilities. Later Lissauer, Cuzzi, and Stanton Peale of the University of California at Santa Barbara suggested that Janus (the larger coorbital), which is coupled to the A ring through resonance excitation of spiral density waves, may have been linked to Enceladus through the latter's 2 : 1 resonance, only to have been torn from that resonance lock 15 to 20 million years ago and set drifting outward. The resonance disruption may have followed the collision that created

Figure 9.3
What makes rings? Some think that rings are as old as the solar system, 4.6 billion years; others think that rings are much younger. No one knows for sure. ["Frank and Ernest," drawing by Bob Thaves reprinted by permission, copyright 1980 by NEA, Inc.]

Janus's companion coorbital, or the formation of the present Enceladus-Dione 2 : 1 resonance may have broken it. A now broken lock to Enceladus would be only a partial solution. It leaves unexplained the failure of the A ring to collapse inward. Still, the hypothesis does point up the possibility that resonance links to massive anchors might have formed and broken from time to time as a satellite moving outward under the influence of the rings drifted into resonances that hold back the rings as well as the satellite.

The Uranian rings present similar problems. A minimum of 10 shepherds must keep the rings from spreading—as well as maintain their inclinations and eccentricities—but there the chain would stop. If the shepherd orbits are located at resonances with the major satellites, or linked to them through resonances with yet undiscovered satellites, the Uranian rings may well have been stable over the lifetime of the solar system. *Voyager 2* might help by locating some shepherds, other outer satellites that might serve to anchor the shepherds, or both. At Jupiter, the particles of the visible ring are not stable in the face of processes that would destroy them or cause them to spiral into Jupiter. The unseen source particles might be herded by the embedded moonlets Metis and Adrastea, but a source of long-term stability is not obvious.

An alternative to a firm foundation for ring systems would be the assumption that they are young and unstable, just too recently formed to show obvious signs of age. Most scientists find this idea philosophically repugnant. They think it unlikely that by chance they live in a special time of abundant rings. They resisted this temptation when Jupiter's ring seemed so young, and they believe that this easy way out will not be the answer to what makes other rings.

10

On to Neptune

After the number of known ringed planets jumped from one to three in 2 years, the question on everyone's mind was, "Are there other ring systems to be found?" In the excitement of discovery, the question sometimes became not whether but where the next ring discovery would be. Most eyes were turning toward Neptune. It was the last of the gas giants with no known rings.

The inner, rocky planets certainly looked unpromising, since they could be closely scrutinized with ground-based telescopes and no rings were to be seen. The fierce solar radiation and strong solar tides there would only have accelerated the dissipation of any rings that might have formed. Mercury and Venus do not have even a single satellite to anchor rings. Mars has only its two tiny satellites, Phobos and Diemos, which may be late arrivals from the asteroid belt rather than original members of the Martian system. Phobos's tidal interaction with Mars is rapidly shrinking its orbit so that in about 100 million years it will crash into the planet. The internal strength of Phobos may keep it in one piece until it hits, but once inside the Roche limit, small particles will leave the surface and form a ring of small particles, probably a short-lived one.

It occurred to Ken Brecher of Boston University and his colleagues that if rings form in a planet's nebula, perhaps a ring formed about the Sun in its nebula and survived until today. It would have to be a tough ring. Its particles would have to be made of material difficult to sublimate—like graphite—be at least 20 kilometers in diameter, and stay at least 4 solar radii from the Sun. No one has run across such a ring, and at least one intentional search has failed to detect a solar ring.

Looking toward the far outer solar system, the prospects for rings looked much better, but there was a nagging problem. Neptune is a gas giant of about the same size and mass as Uranus, but there is a potentially crucial oddness about it. The major satellites of Jupiter, Saturn, and Uranus all orbit in the same direction that their planets rotate, just as all the planets orbit the Sun in the direction of its rotation. The major satellites of the three gas giants and eight of the nine planets orbit near the equatorial planes of their central bodies in nearly circular orbits. Formation from a single rotating nebular disk presumably imposes such regularity on the solar system and planetary systems.

The Neptune system is different. One of its two confirmed satellites is Triton, which has a circular orbit inclined 20 degrees to Neptune's equator. Triton revolves in a *retrograde* direction—that is, "backward" with respect to the rotation and orbital revolution of almost all other objects in the solar system. The orbit of Neptune's smaller satellite, Nereid, is both wildly eccentric and steeply inclined. Its orbit is so large, though, that Neptune probably captured it, as Jupiter and Saturn appear to have captured some of their small, outer satellites.

Because Triton's retrograde motion does not fit naturally into conventional theories of planetary origins, some have suggested more catastrophic explanations. Perhaps a massive intruder, now a tenth planet lurking somewhere unseen beyond the known edge of the solar system, plunged through the Neptune system and disrupted the satellite orbits. Considering Pluto's small size (smaller than our moon) and its highly inclined (17 degrees) and eccentric orbit, perhaps it was a third satellite of Neptune that the intruder threw out of the system into an orbit around the Sun. Many theorists have questioned such scenarios on several grounds, especially since the 1978 discovery of Pluto's relatively gigantic moon, Charon.

Whatever made the Neptune and Pluto systems so peculiar began to bother those figuring the odds on the discovery of new ring systems. Perhaps rings only form and persist within regular satellite systems like those around Jupiter, Saturn, and Uranus. If Triton and Neptune formed together, perhaps rings never formed. If they did form, they may have been subsequently destroyed. Or another retrograde satellite spiraled into the planet, as Triton may be ever so slowly doing now, and swept up the ring before the satellite perished in Neptune's atmosphere. Considering the murkiness of theorizing on ring origins, the obvious best course was direct observation. If Neptune or even Pluto had rings, that would imply that rings are a relatively robust phenomenon, at least in the outer solar system. If Neptune, the prime candidate, had no rings, it would argue in favor of some special relation between rings and the regular satellite systems of the gas giants.

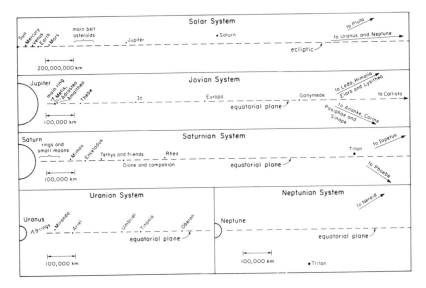

Figure 10.1

Do "regular" satellites and rings go together? These views look along the equatorial planes of Jupiter, Saturn, Uranus, and Neptune. For comparison with the satellite systems, a view along the ecliptic (the plane of the Earth's orbit) is shown for the solar system. On the compressed scale needed to contain the planetary orbits, even the Sun appears only as a point, and the solar rotation axis is about 7 degrees from the perpendicular to the ecliptic. Each satellite with an orbital plane inclined with respect to the equatorial plane of the central planet has been plotted at its angle of inclination, above the plane if its orbital motion is in the same sense as the planet's rotation and below the plane if its orbital motion is in the opposite sense. Each of the three planets known to have rings—Jupiter, Saturn, and Uranus—has a family of several satellites in nearly equatorial orbits. These satellites probably formed along with the planet. The distant satellites with inclined orbits are probably stray asteroids that were captured later. In contrast with these three "regular" satellite systems, Neptune has only one established nearby satellite, Triton, which has an orbit inclined with respect to Neptune's equator and revolves in the opposite sense to Neptune's rotation. So far, no rings have been detected around Neptune, suggesting a connection between ring systems and regular satellites.

Astronomers had been searching for Neptunian rings in an indirect sort of way during satellite searches since the planet's discovery in 1846—recall Lassell's unfounded report of a ring 2 weeks later. If Neptune had rings, they were not easily seen, even with the most powerful visual telescopes under the most favorable observing conditions. Astronomers needed more powerful techniques to search for rings. They would have it without knowing it.

When observers in Australia, New Zealand, and Japan recorded an occultation by Neptune in 1968, they had no thought of rings, although the occultation technique was many times more sensitive for detecting dark, narrow rings than visual observation or photography. As it turned out, none of the published records covered the critical ring region. Once rings came into vogue, Ken Freeman, the principal observer in Australia, reported privately to colleagues that, although he had discarded data from the zone where rings would be expected, he had checked all that data and would have noticed anything obvious. On this basis, astronomers dismissed the possibility of an extensive ring system located at the usual distance from the planet. But a complication arose. They still do not know where Neptune's pole is to within a few degrees, so that the position of equatorial rings remains slightly uncertain. Furthermore, Triton's inclined orbit would allow the presence of stable, warped rings that could lie outside of the equatorial plane. Rings could even be stable running over the poles. Hence, the lack of ring evidence from the 1968 occultation only slightly dampened the enthusiastic spirits of those eager to search for rings around Neptune.

As with Uranus, Neptune would be expected to occult stars not contained within the SAO catalog that would still be suitable events for ring searches. In the fall of 1977, Arnold Klemola, Brian Marsden, Bill Liller, and I established a list of stars that Neptune would probably occult during the next few years. However, this list contained fewer exceptional candidates than did the corresponding list for Uranus, since Neptune's rings would subtend only about 2 arc-seconds perpendicular to the direction of Neptune's motion compared with the 8 arc-seconds subtended by the ϵ ring of Uranus. The brightest star on the list would be occulted on 10 February 1980 and should have been visible from an area of the Pacific Ocean near Hawaii. Neptune would just graze the star, but that would probe an extensive region for rings. With the stable photometry possible from the KAO, we could probe for rings with significantly greater sensitivity than the 1968 occultation achieved. We made a special appeal to NASA headquarters to approve these observations, which would require finding funds to support the work as well as rearranging the schedule of the KAO. The reception was generally positive. But while headquarters was trying to accommodate our request, I found an effect that we had forgotten to include in our calculation for the occultation zone. The correct calculation showed the star to be farther away from Neptune, so that the region probed would not be close enough to the planet to be interesting—or at least interesting enough to justify a large airborne expedition. Just in case the prediction was wrong, Bob Millis and I observed the event with the Lowell planetary patrol telescope on Mauna Kea. The skies were clear and the result was as expected: no evidence for rings, but Neptune apparently passed too far from the star to probe the most likely region for rings.

Phil Nicholson and Terry Jones seized the next opportunity, an event observable from Mount Stromlo in 1980. Conditions were not perfect, but they acquired reasonably good data. Their signal dipped a couple of times; but without confirming evidence, they could conclude nothing.

In 1981, Neptune would occult a bright star, an event that would be visible from sites in the western and central Pacific Ocean. Since the Uranian rings were discovered in 1977 and the Jovian ring in 1979, the next year in the sequence should be 2 years later, 1981—the year for discovering Neptune's rings. I set up an extensive observing network at Mauna Kea, Mount Stromlo, and Siding Spring. Observers monitored the occultation at several telescopes and at several wavelengths simultaneously at each site. Thus we could subject any candidate ring events to careful scrutiny. In addition, we identified a second occultation—by a star that we had previously thought would be too faint—that would be visible from Cerro Tololo 2 weeks later. Fortunately, Jay Elias was scheduled to be observing on the 4-meter telescope and was interested in searching for rings during the occultation. He had excellent conditions and he probed within 6,000 kilometers from the top of Neptune's atmosphere. One "glitch" in the signal could be ruled out as a ring event because it would have been too narrow a shadow at Earth to be a real physical object at Neptune, unless it was a satellite that happened to be moving at the right speed and in the right direction. We said that we could find no rings 5 kilometers or wider that were more opaque than an optical depth of 0.07 at distances greater than 31,400 kilometers from the center of Neptune (whose radius is about 25,200 kilometers). That meant that we would have easily detected ring systems as extensive as those of Saturn or Uranus, but not one as tenuous as the Jovian system. Of course, a ring system as extensive as that of Saturn could have been easily seen in a telescope.

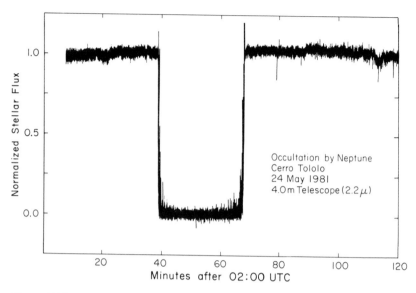

Figure 10.2

No dips this time. These data, recorded by Jay Elias at Cerro Tololo, show an occultation by Neptune, but no occultations by rings. The first dip after the Neptune occultation, upon closer inspection at higher resolution, proved to be too narrow to have been caused by a ring—a ring shadow would have been broadened to about 5 kilometers by diffraction effects. Comparing these data with the Uranian ring occultation data in figure 4.3, we conclude that any ring system of Neptune must be much less extensive than that of Uranus. However, a tenuous ring, similar to Jupiter's, cannot be ruled out. More sensitive searches can be carried out after the launch of the Space Telescope (currently scheduled for the summer of 1986) and during the Voyager encounter in August 1989 (if the spacecraft still functions well). [Reprinted, by permission, from J. L. Elliot, D. J. Mink, J. H. Elias, R. L. Baron, E. Dunham, J. E. Pingree, R. G. French, W. Liller, P. D. Nicholson, T. J. Jones, and O. G. Franz, "No Evidence of Rings around Neptune," *Nature* 294 (1981), 526, copyright 1981 by Macmillan Journals Limited]

The next two events in the Neptune saga seemed to increase the likelihood that the planet would have rings. The first occurred in 1981, when Harold Reitsema and his colleagues at the University of Arizona apparently detected a third satellite of Neptune while monitoring the light from a star as Neptune passed close by it. With two telescopes, each equipped with two-color photometers, they recorded simultaneous dips in their red channels but not in the blue. Something seemed to block the reddish star, but not the bluish planet. Their interpretation that they detected a third satellite is likely, but not certain. If they are correct, the occultation would have been highly improbable unless Neptune has numerous small satellites in nearby orbits. Elsewhere, small inner satellites confine, shape, and even rejuvenate rings. Why not at Neptune?

At the spring meeting of the American Astronomical Society in 1982 at Troy, New York, Ed Guinan and his colleagues reported that some occultation data that Guinan had recorded in New Zealand during the 1968 occultation, but only recently analyzed, showed a large dip of about 30 percent lasting 2 minutes and 48 seconds. If a ring caused this dip, it would be inconsistent with Ken Freeman's recollection of "nothing obvious" in his data. Furthermore, the possible ring only barely avoided the area ruled out by the 1981 observations. One would have to assume an extreme position for the pole of Neptune for our 1981 results to be consistent with his purported ring. Although the *New York Times* on 10 June reported "DATA SHOW TWO RINGS CIRCLING NEPTUNE: Astronomer Says They Appear to Be 1,200 Miles Wide—Makeup Undetermined," most specialists in the field remained skeptical. We wanted to believe that Neptune had a ring system. However, at best Guinan's results had no corroboration; more likely they conflicted with the negative evidence of two previous occultations.

The next occultation search came in 1983. It was a bright star, and the geometry was such that at the proper location on Earth we could probe virtually to the top of Neptune's atmosphere. Several groups made extensive preparations to observe from sites in Australia, New Zealand, Japan, Indonesia, Taiwan, Guam, and Hawaii. With the KAO, we could get to the best geometrical location. This appeared to be near Australia at the beginning, but more measurements of the star's position in the spring by Arnold Klemola showed that the best position would be closer to Guam's Andersen Air Force Base. The KAO flight proved exciting; we had to change our course northward during the flight to avoid cirrus cloud above us. Visible and infrared satellite pictures received at Guam defined the cloud system well; being clouded

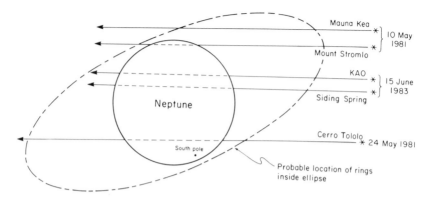

Figure 10.3

Scanning for rings. Several occultations of stars by Neptune have been observed in recent years, and the lines in the figure show the paths of the occulted stars through the Neptune system. See figure 10.2 for a sample of the results.

out from the KAO would have been difficult to explain and downright embarrassing.

The 1983 search was even more exhaustive, but still no rings. Guinan's possible ring could not exist. It must have been a passing cloud or the result of a tracking error in the telescope. This probably marks the most extensive occultation search for rings that will be performed with ground-based occultations for some time. More sophisticated imaging of Neptune in the infrared, where Neptune is relatively dark, by David Allen with the Anglo-Australian Telescope at Siding Spring revealed no rings either. If Neptune has rings, they almost certainly will not be discovered from the ground.

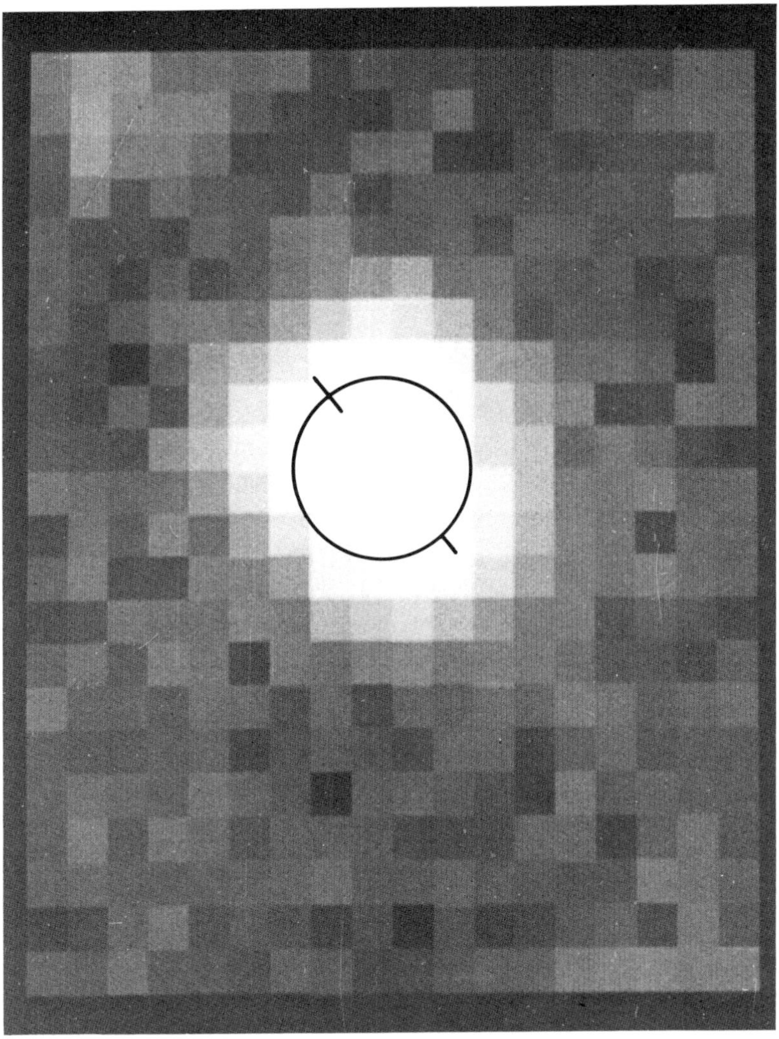

Figure 10.4
Where are the rings? This infrared image of Neptune taken at a wavelength of 2.2 microns by David Allen with the Anglo-Australian Telescope at Siding Spring shows no rings. The outline of the planet and the orientation of its pole, shown by the solid lines, have been added to the photograph. Each pixel corresponds to about 11,000 kilometers at the distance of Neptune, and the resolution of the image is about 22,000 kilometers. [Courtesy of D. Allen; copyright 1983 by Anglo-Australian Telescope Board]

11

Future Rings

Galileo started it. He turned his telescope on Saturn and improved man's view by a factor of 30. That allowed him to be the first to see Saturn's rings. Then, using an improved telescope, Cassini discovered the gap between the two major rings. Much later, stellar occultations improved resolution about 10,000 times over that of modern ground-based imaging, leading to the discovery of the Uranian rings. A similar improvement for Saturn by spacecraft imaging revealed the myriad ring features there. Finally, the combination of spacecraft transportation and stellar occultation observations produced the highest resolution yet, and, not surprisingly, the most interesting ring detail. The future promises further increases in resolving power and light sensitivity. New and exciting discoveries seem inevitable.

Some of the more modest resolution improvements will come as ground-based observations of occultations continue to accumulate, improving measurements of the sizes and shapes of the Uranian rings. These should reveal ever more subtle dynamical effects, such as twisting and warping, that could unequivocally demand the presence of Uranian shepherds. In 1982, in fact, Dick French and I used our ever increasing set of occultation data to establish that some of the Uranian rings are inclined with respect to the equatorial plane of Uranus by a few hundredths of a degree. Although rings should not be inclined according to Jeffreys's arguments, Borderies, Goldreich, and Tremaine have extended their shepherd model to explain the inclinations, as well as the eccentricities and sharp edges of the Uranian rings. Surprises are always possible whenever a ring system occults a brighter star of smaller angular diameter, giving thereby a more sensitive probe of their structure. And there is always the chance of a lucky detection of a small satellite, such as Reitsema's apparent Neptunian satellite, that might prove significant for understanding ring structure.

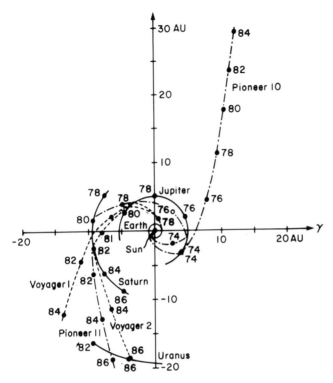

Figure 11.1
Paths of exploration. During the past decade, two Pioneer and two Voyager spacecraft have blazed trails to the outer solar system. Positions are as of January of the years indicated.

Spacecraft will be providing even better views of some rings and searching for new rings. Once the Galileo spacecraft starts orbiting Jupiter in 1988, it will have a more leisurely look at the Jovian ring than Voyager did, making it possible to locate more precisely the ring's two companions. Whether Galileo returns sharper images of the ring will depend on the exact orbits chosen for it; like Voyager, it will be fighting the smearing effect of spacecraft motion. Galileo will be imaging the ring at a number of wavelengths and at a variety of angles with respect to the Sun. That should provide more information on particle size and composition.

The *Voyager 2* encounter with Uranus during January 1986 should provide greatly increased resolution of the rings, but it will not be all that easy. The imaging team will have to pick its shots carefully. Pitch black and 2.9 billion kilometers from the Sun, Uranian ring particles will be a difficult target for the approaching Voyager. Once past the planet, it may have an easier time of it in forward-scattered light, as happened at Jupiter. Imaging's greatest achievement would be the detection of the predicted shepherd satellites. Detection of satellites between Miranda and the rings may also reveal particle sources and additional generators of resonances. Another key observation for Voyager would be a search for yet undetected rings and particles between the nine narrow rings. Particles smaller than the size that shepherds can efficiently herd should leak out of the rings and spiral closer to Uranus under the influence of the Poynting-Robertson effect. By this reasoning, a tenuous yet undetected sheet of fine particles should exist between the rings. Its detection and the determination of the particle sizes involved would reveal the strength of the shepherding force— or whatever the mechanism is producing the sharp edges of the rings. As at Saturn, the most astounding increase in resolution will come from the photopolarimeter's observation of one (or more with any luck) stellar occultation.

Before *Voyager 2* has a chance to reach Neptune, the Space Telescope should be operating in Earth orbit. Scheduled for launch in 1986, the Space Telescope will resolve detail about 10 times better than achievable from the ground. It will also have the advantage of a lower background of scattered light, permitting more sensitive searches for rings around Neptune and Pluto. Observations of known ring systems will benefit as well; for example, Space Telescope will follow Saturn's B ring spokes as they sweep around the planet. It also offers improved capability for occultation observations, including occultations by Saturn's rings. Such occultations would have a resolution of a few kilometers, not far above Voyager's best imaging resolution. And they could be repeated. Space Telescope could also use occultations to detect the diffuse ring of Jupiter and to search for other faint rings. If Space Telescope fails to find any Neptunian rings, the next search will be up to Voyager when it arrives in August 1989. If the spacecraft is still functioning properly, it should make what will probably be the most sensitive search for rings for a long time to come.

No one ever accused NASA of not thinking big, and rings are no exception. A Saturn ring mission described in a NASA concept development study would provide the ultimate in resolution. Raw power would be the key. A 100-kilowatt nuclear electric propulsion system would slowly spiral the spacecraft away from Earth and decelerate it on arrival at Saturn until it reached the outer edge of the A ring. From there, continuous thrusting by the same engine would lift it just a few kilometers above the ring plane and begin its 10-month trip across the rings, like a stylus of a phonograph gliding across a record. If it ever encountered a thin spot or a true gap, it could cut its engines and fall through the rings or even thrust for a ring particle rendezvous. Instead of Voyager's mad dash through the ring plane, this mission could linger among the ring particles in order to extract their secrets of shape, size, composition, strength, and collision behavior.

The next decade of ring studies seems to be well planned. Astronomers can expect a steady increase in sensitivity and resolution that is sure to reveal new details of the rings and provide new insight to their workings. But we must hope that all does not go exactly according to plan. Science without serendipity would not be sterile, but it would be measurably less rewarding, less exciting. Who can imagine all that the rings have in store?

Figure 11.2
What does a ring particle look like? The exact sizes and shapes of ring particles will remain a mystery until we can obtain close-up imaging, such as would be possible with the ring-encounter mission illustrated. [Courtesy of NASA]

These ring maps show the main structural features and major satellites of the three known ringed planets out to a distance of 4 planetary radii. For comparison, a kilometer scale is also given. In each figure the synchronous orbit, for which an orbiting body has the same revolution period as the planet's rotation period, has been indicated, along with the positions of the Roche limit for material of densities 1.0 and 2.0 grams per cubic centimeter.

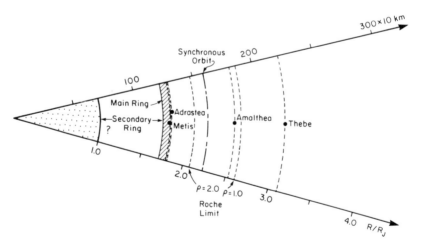

Jupiter's ethereal ring (pie slice).

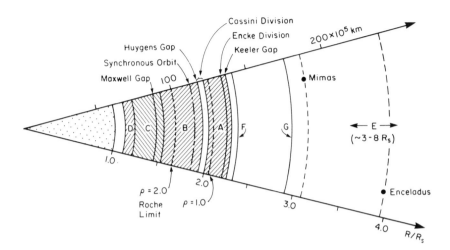

See it all at Saturn (pie slice).

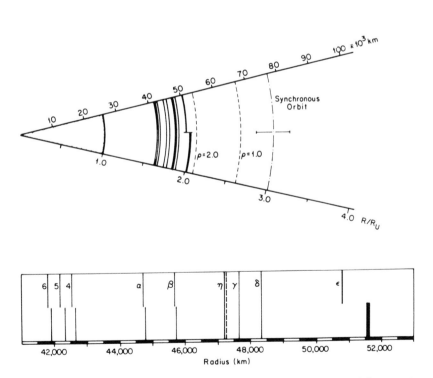

The Uranian bull's-eye (pie slice). [Courtesy of P. Nicholson; reprinted, by permission, from J. L. Elliot and P. D. Nicholson, "The Rings of Uranus," in *Planetary Rings*, edited by A. Brahic and R. Greenberg (Tucson: University of Arizona Press, 1984)]

Bibliography

Chapter 1

Elliot, James L., Edward Dunham, and Robert L. Millis. "Discovering the Rings of Uranus." *Sky and Telescope* 53 (1977), 412–416, 430.

Chapter 2

Alexander, A. F. O'D. *The Planet Saturn: A History of Observation, Theory and Discovery.* London: Faber and Faber, 1962.

Bobrov, M. S. "Physical Properties of Saturn's Rings." In *Surfaces and Interiors of Planets and Satellites,* edited by A. Dollfus. New York: Academic Press, 1970.

Burke, T., C. Hord, C. Lillie, A. Cook, J. Bertaux, and G. L. Tyler. "Report of the Saturn's Rings Working Group." 8 April 1974. Prepared for Edward Stone.

"The Composition of Saturn's Rings." *Sky and Telescope* 39 (1970), 14.

Cook, Allan, and Fred Franklin. "Saturn's Rings: A New Survey." In *Planetary Satellites,* edited by J. Burns. Tucson: University of Arizona Press, 1977.

Cook, A. F., F. A. Franklin, and F. D. Palluconi. "Saturn's Rings—a Survey." *Icarus* 18 (1973), 317–337.

Cuzzi, Jeffrey. "The Rings of Saturn: State of Current Knowledge and Some Suggestions for Future Studies." In *The Saturn System,* edited by Donald Hunten and David Morrison. NASA Conference Publication 2068, 1978.

Cuzzi, Jeffrey, and James Pollack. "Saturn's Rings: Particle Composition and Size Distribution as Constrained by Microwave Observations. I. Radar Observations." *Icarus* 33 (1978), 233–262.

Cuzzi, Jeffrey, James Pollack, and Audrey Summers. "Saturn's Rings: Particle Composition and Size Distribution as Constrained by Observations at Microwave Wavelengths. II. Radio Interferometric Observations." *Icarus* 44 (1980), 683–705.

Franklin, F. A., G. Colombo, and A. F. Cook. "A Dynamical Model for the Radial Structure of Saturn's Rings." *Icarus* 15 (1971), 80–92.

Kawata, Y., and W. M. Irvine. "Models of Saturn's Rings Which Satisfy the Optical Observations." In *Exploration of the Planetary System,* edited by Woszcyzyk and Iwaniszewska. International Astronomical Union, 1974.

Kuiper, Gerard P., Dale P. Cruikshank, and Uwe Fink. Letter in *Sky and Telescope* 39 (1970), 80.

Lumme, Kari, and William Irvine. "A Model for the Azimuthal Brightness Variations in Saturn's Rings." *Nature* 282 (1979), 695–696.

Lumme, Kari, William Irvine, and Larry Esposito. "Theoretical Interpretation of the Ground Based Photometry of Saturn's Rings." *Icarus* 53 (1983), 174–184.

Osterbrock, Donald, and Dale Cruikshank. "J. E. Keeler's Discovery of a Gap in the Outer Part of the A Ring." *Icarus* 53 (1983), 165–173.

Palluconi, Frank, and Gordon Pettengill, eds. *The Rings of Saturn.* Proceedings of the Saturn's Rings Workshop, 31 July and 1 August 1973. NASA SP-343. Washington, DC: National Aeronautics and Space Administration, 1974.

Pilcher, Carl B., Clark R. Chapman, Larry A. Lebofsky, and Hugh H. Kieffer. "Saturn's Rings: Identification of Water Frost." *Science* 167 (1970), 1372–1373.

Pollack, James. "The Rings of Saturn." *Space Science Reviews* 18 (1975), 3–93.

Pollack, James. "The Rings of Saturn." *American Scientist* 66 (1978), 30–37.

Reitsema, H. J. "Photometric Confirmation of the Encke Division in Saturn's A Ring." *Nature* 272 (1978), 601–602.

Van Helden, Albert. "Saturn through the Telescope: A Brief Historical Survey." In *Saturn*, edited by T. Gehrels and M. Matthews. Tucson: University of Arizona Press (1984).

Van Helden, Albert. "Rings in Astronomy and Cosmology, 1600–1900." In *Planetary Rings*, edited by R. Greenberg and A. Brahic. Tucson: University of Arizona Press (1984).

Chapter 3

Baum, W. A., and A. D. Code. "A Photometric Observation of the Occultation of σ Arietis by Jupiter." *Astronomical Journal* 58 (1953), 108–112.

Brinkman, R. T. "Occultation by Jupiter." *Nature* 230 (1971), 515–516.

Combes, M., J. Lecacheux, and L. Vapillon. "First Results of the Occultation of β Sco by Jupiter." *Astronomy and Astrophysics* 15 (1971), 235–238.

Combes, M., L. Vapillon, and J. Lecacheux. "The Occultation of β Scorpii by Jupiter. IV. Divergences with Other Observers in the Derived Temperature Profiles." *Astronomy and Astrophysics* 45 (1975), 399–403.

De Vaucouleurs, G., and D. H. Menzel. "Results of the Occultation of Regulus by Venus, July 7, 1959." *Nature* 188 (1960), 28–33.

Elliot, James L. "Signal-to-Noise Ratios for Occultations by the Rings of Uranus 1977–1980." *Astronomical Journal* 82 (1977), 1036–1038.

Elliot, James L. "Stellar Occultation Studies of the Solar System." *Annual Reviews of Astronomy and Astrophysics* 17 (1979), 445–475.

Elliot, J. L., E. Dunham, and C. Church. "A Unique Airborne Observation." *Sky and Telescope* 52 (1976), 23–25.

Elliot, J. L., K. Rages, and J. Veverka. "The Occultation of Beta Scorpii by Jupiter. VI. The Masses of Beta Scorpii A_1 and A_2." *Astrophysical Journal (Letters)* (1975), L123–L126.

Elliot, J. L., J. Veverka, and R. L. Millis. "Uranus Occults SAO 158687." *Nature* 265 (1977), 609–611.

Elliot, J. L., L. H. Wasserman, J. Veverka, Carl Sagan, and W. Liller. "The Occultation of Beta Scorpii by Jupiter. II. The Hydrogen-Helium Abundance in the Jovian Atmosphere." *Astrophysical Journal* 190 (1974), 719–729.

Elliot, J. L., R. G. French, E. Dunham, P. Gierasch, J. Veverka, C. Church, and Carl Sagan. "Occultation of ε Geminorum by Mars: Evidence for Atmospheric Tides?" *Science* 195 (1977), 485–486.

Elliot, J. L., R. G. French, E. Dunham, P. Gierasch, J. Veverka, C. Church, and Carl Sagan. "Occultation of ε Geminorum by Mars. II. The Structure and Extinction of the Martian Upper Atmosphere." *Astrophysical Journal* 217 (1977), 661–679.

Freeman, K. C., and G. Lyngå. "Data for Neptune from Occultation Observations." *Astrophysical Journal* 160 (1970), 767–780.

Hubbard, W. B., R. E. Nather, David S. Evans, R. G. Tull, D. C. Wells, G. W. van Citters, B. Warner, and P. Vanden Bout. "The Occultation of Beta Scorpii by Jupiter and Io. I. Jupiter." *Astronomical Journal* 77 (1972), 41–59.

Liller, W., J. L. Elliot, J. Veverka, L. H. Wasserman, and C. Sagan. "The Occultation of β Scorpii III. Simultaneous High Time-Resolution Records at Three Wavelengths." *Icarus* 22 (1974), 82–104.

Martynov, D. Ya. "The Radius of Venus." *Soviet Astronomy* 4 (1961), 798–804.

Osawa, Kiyoteru, Kihachiro Ichimura, and Minoru Shimizu. "Occultation of BD −17°4388 by Neptune on 7 April 1968 (II), Observation at Okayama Station and the Scale Height of Neptune." *Tokyo Astronomical Bulletin* (2nd Ser.) 184 (1968), 2183–2187.

Pannekoek, Ant. "Über die Erscheinungen, welche bei einer Sternbedeckung durch einen Planeten auftreten." *Astronomische Nachrichten* 164 No. 3913 (1903), 5–10.

Taylor, Gordon E. "An Occultation by Uranus." *Journal of the British Astronomical Association* 83 (1973), 352.

Veverka, Joseph, and Lawrence Wasserman. "The Regulus Occultation and the Real Atmosphere of Venus." *Icarus* 21 (1974), 196–198.

Veverka, J., L. H. Wasserman, J. Elliot, and Carl Sagan. "The Occultation of β Scorpii by Jupiter. I. The Structure of the Jovian Upper Atmosphere." *Astronomical Journal* 79 (1974), 73–84.

Chapter 4

"Belt of Satellites Discovered around Uranus." *Science News* 111 (1977), 180.

Bhattacharyya, J. C., and M. K. V. Bappu. "Saturn-Like Ring System around Uranus." *Nature* 270 (1977), 503–506.

Bhattacharyya, J. C., and K. Kuppuswamy. "A New Satellite of Uranus." *Nature* 267 (1977), 331–332.

Bhattacharyya, J. C., M. K. V. Bappu, S. Mohin, H. S. Mahra, and S. K. Gupta. "Extended Ring System of Uranus." *The Moon and the Planets* 21 (1979), 393–404.

Boynton, Judd. "Saturn's Rings Are in Orbital Spiral." *Proceedings, Joint Conference on Astronomy* (1975), 14–15.

Chen Dao-han, Yang Hsiu-yi, Wu Chih-hsien, Wu Yueh-chen, Kiang Shih-yang, Huang Yung-wei, Yeh Chi-tang, Chai Ti-sheng, Hsieh Chung-cheih, Cheng Chien-sheng, and Chang Chin. "Photoelectric Observation of SAO 158687 by Uranian Ring and the Detection of Uranian Ring Signals from the Light Curve." *Scientia Sinica* XXI (1978), 503–508, and Plates I, II, and III.

De, Bibhas R. "A 1972 Prediction of the Uranian Rings Based on the Alfvén Critical Velocity Effect." *The Moon and the Planets* 18 (1978), 339–342.

Elliot, J. L., and P. D. Nicholson. "The Rings of Uranus." In *Planetary Rings*, edited by A. Brahic and R. Greenberg. Tucson: University of Arizona Press, 1984.

Elliot, J. L., E. Dunham, and D. Mink. "The Rings of Uranus." *Nature* 267 (1977), 328–330.

Elliot, J. L., E. Dunham, L. H. Wasserman, R. L. Millis, and J. Churms. "The Radii of Uranian Rings α, β, γ, δ, ϵ, η, 4, 5, and 6 from Their Occultations of SAO 158687." *Astronomical Journal* 83 (1978), 980–992.

Elliot, J. L., R. G. French, J. A. Frogel, J. H. Elias, D. J. Mink, and W. Liller. "Orbits of Nine Uranian Rings." *Astronomical Journal* 86 (1981), 444–455.

Elliot, J. L., Jay A. Frogel, J. H. Elias, I. S. Glass, R. G. French, D. J. Mink, and W. Liller. "The 20 March 1980 Occultation by the Uranian Rings." *Astronomical Journal* 86 (1981), 127–134.

French, R. G., J. L. Elliot, and D. A. Allen. "Inclinations of the Uranian Rings." *Nature* 298 (1982), 827–829.

Goldreich, Peter, and Scott Tremaine. "Toward a Theory for the Uranian Rings." *Nature* 277 (1979), 97–99.

Hubbard, W. B., G. V. Coyne, T. Gehrels, B. A. Smith, and B. H. Zellner. "Observations of Uranus Occultation Events." *Nature* 268 (1977), 33–34.

McLaughlin, William I. "Prediscovery Evidence of Planetary Rings." *Journal of the British Interplanetary Society* 33 (1980), 287–294.

Millis, R. L., and L. H. Wasserman. "The Occultation of BD-15°3969 by the Rings of Uranus." *Astronomical Journal* 83 (1978), 993–998.

Millis, R. L., L. H. Wasserman, and P. V. Birch. "Detection of Rings around Uranus." *Nature* 267 (1977), 330–331.

Nicholson, P. D., and T. J. Jones. "Two-Micron Spectrophotometry of Uranus and Its Rings." *Icarus* 42 (1980), 54–67.

Nicholson, P. D., K. Matthews, and P. Goldreich. "The Uranus Occultation of 10 June 1979. I. The Rings." *Astronomical Journal* 86 (1981), 596–606.

Nicholson, P. D., K. Matthews, and P. Goldreich. "Radial Widths, Optical Depths, and Eccentricities of the Uranian Rings." *Astronomical Journal* 87 (1982), 433–447.

Nicholson, P. D., S. E. Persson, K. Matthews, P. Goldreich, and G. Neugebauer. "The Rings of Uranus: Results of the 10 April 1978 Occultation." *Astronomical Journal* 83 (1978), 1240–1248.

Sicardy, B., M. Combes, A. Brahic, P. Bouchet, C. Perrier, and R. Courtin. "The 15 August 1980 Occultation by the Uranian System: Structure of the Rings and Temperature of the Upper Atmosphere." *Icarus* 52 (1982), 454–472.

Sinton, W. M. "Uranus: The Rings Are Black." *Science* 198 (1977), 503–504.

Tomita, Koichiro. "Observation of Occultation of the SAO 158687 Star by Uranus at Dodaira Station." *Tokyo Astronomical Bulletin* (2nd Ser.) 250 (1977), 2885–2888.

Chapter 5

Acuña, Mario, and Norman F. Ness. "The Main Magnetic Field of Jupiter." *Journal of Geophysical Research* 81 (1976), 2917–2922.

Becklin, E. E., and C. G. Wynn-Williams. "Detection of Jupiter's Ring at 2.2 μ." *Nature* 279 (1979), 400–401.

Burns, Joseph A., Mark R. Showalter, Jeffrey N. Cuzzi, and James B. Pollack. "Physical Processes in Jupiter's Ring: Clues to Its Origin by Jove!" *Icarus* 44 (1980), 339–360.

Fillius, Walker. "The Trapped Radiation Belts of Jupiter." In *Jupiter*, edited by T. Gehrels. Tucson: University of Arizona Press, 1976.

Fillius, R. Walker, Carl E. McIlwain, and Antonio Mogro-Campera. "Radiation Belts of Jupiter: A Second Look." *Science* 188 (1975), 465–467.

Humes, D. H. "The Jovian Meteoroid Environment." In *Jupiter*, edited by T. Gehrels. Tucson: University of Arizona Press, 1976, 1052–1067.

Humes, D. H. "Results of Pioneer 10 and 11 Meteoroid Experiments: Interplanetary and Near-Saturn." *Journal of Geophysical Research* 85 (1980), 5841–5852.

Humes, D. H., J. M. Alvarez, W. H. Kinard, and R. L. O'Neal. "Pioneer 11 Meteoroid Detection Experiment: Preliminary Results." *Science* 188 (1975), 473–474.

Humes, D. H., J. M. Alvarez, R. L. O'Neal, and W. H. Kinard. "The Interplanetary and Near-Jupiter Meteoroid Environments." *Journal of Geophysical Research* 79 (1974), 3677–3684.

Ip, W.-H. "On the Pioneer 11 Observation of the Ring of Jupiter." *Nature* 280 (1979), 478–479.

Jewitt, D. C., and G. E. Danielson. "The Jovian Ring." *Journal of Geophysical Research* 86 (1981), 8691–8697.

Jewitt, D. C., G. E. Danielson, and R. J. Terrile. "Ground-Based Observations of the Jovian Ring and Inner Satellites." *Icarus* 48 (1981), 536–539.

Kinard, W. H., R. L. O'Neal, J. M. Alvarez, and D. H. Humes. "Interplanetary and Near-Jupiter Meteoroid Environments: Preliminary Results from the Meteoroid Detection Experiment." *Science* 183 (1974), 321–322.

Neugebauer, G., E. E. Becklin, D. C. Jewitt, R. J. Terrile, and G. E. Danielson. "Spectra of the Jovian Ring and Amalthea." *Astronomical Journal* 86 (1981), 607–610.

Owen, Tobias, G. Edward Danielson, Allan F. Cook, Candice Hansen, Virginia L. Hall, and Thomas Duxbury. "Jupiter's Rings." *Nature* 281 (1979), 442–446.

Roederer, Juan G., Mario H. Acuña, and Norman F. Ness. "Jupiter's Internal Magnetic Field Geometry Relevant to Particle Trapping." *Journal of Geophysical Research* 82 (1977), 5187–5194.

Smith, Bradford A., et al. "The Jupiter System through the Eyes of Voyager." *Science* 204 (1979), 951–972.

Smith, Bradford A., et al. "The Galilean Satellites and Jupiter: Voyager 2 Imaging Science Results." *Science* 206 (1979), 927–950.

Smoluchowski, R. "The Ring Systems of Jupiter, Saturn, and Uranus." *Nature* 280 (1979), 377–378.

Chapter 6

Dallas, S. S., et al. *The D-Ring—Fact or Fiction.* Document 760-134. Pasadena: Jet Propulsion Laboratory, 1975.

Esposito, Larry W., James P. Dilley, and John W. Fountain. "Photometry and Polarimetry of Saturn's Rings from Pioneer Saturn." *Journal of Geophysical Research* 85 (1980), 5948–5956.

Feibelman, W. A. "Concerning the 'D' Ring of Saturn." *Nature* 214 (1967), 793–794.

Feibelman, W. A., and D. A. Klinglesmith III. "Saturn's E Ring Revisited." *Science* 209 (1980), 277–279.

Fillius, W., W. H. Ip, and C. E. McIlwain. "Trapped Radiation Belts of Saturn: First Look." *Science* 207 (1980), 425–431.

Gehrels, Tom, and Larry Esposito. "Pioneer Flyby of Saturn and Its Rings." *Advances in Space Research* 1 (1981), 67–71.

Gehrels, T., et al. "Imaging Photopolarimeter on Pioneer Saturn." *Science* 207 (1980), 434–439.

Guérin, Pierre. "The New Ring of Saturn." *Sky and Telescope* 40 (1970), 88.

Guérin, P. "Les Anneaux de Saturne en 1969." *Icarus* 19 (1973), 202–211.

Humes, D. H. "Results of Pioneer 10 and 11 Meteoroid Experiments: Interplanetary and Near-Saturn." *Journal of Geophysical Research* 85 (1980), 5841–5852.

Humes, D. H., R. L. O'Neal, W. H. Kinard, and J. M. Alvarez. "Impact of Saturn Ring Particles on Pioneer 11." *Science* 207 (1980), 443–444.

Larson, S. M. "Observations of the Saturn D Ring." *Icarus* 37 (1979), 399–403.

National Aeronautics and Space Administration. *Pioneer Saturn Encounter*. Washington, DC: NASA, 1979.

Simpson, J. A., T. S. Bastian, D. L. Chenette, G. A. Lentz, R. B. McKibben, K. R. Pyle, and A. J. Tuzzolino. "Saturnian Trapped Radiation and Its Absorption by Satellites and Rings: The First Results from Pioneer 11." *Science* 207 (1980), 411–415.

Smith, Bradford A. "The D and E Rings of Saturn." In *The Saturn System*, edited by Donald M. Hunten and David Morrison. Washington, DC: National Aeronautics and Space Administration, 1978. (Available from the National Technical Information Service, Springfield, VA 22161.)

Smith, Bradford A., Allan F. Cook II, Walter A. Feibelman, and Reta F. Beebe. "On a Suspected Ring External to the Visible Rings of Saturn." *Icarus* 25 (1975), 466–469.

Van Allen, James A. "Findings on Rings and Inner Satellites of Saturn by Pioneer 11." *Icarus* 51 (1982), 509–527.

Van Allen, J. A., M. F. Thomsen, B. A. Randall, R. L. Rairden, and C. L. Grosskreutz. "Saturn's Magnetosphere, Rings, and Inner Satellites." *Science* 207 (1980), 415–421.

Chapter 7

Burke, T., C. Hord, C. Lillie, A. Cook, J. Bertaux, and G. L. Tyler. *Report of the Saturn's Rings Working Group* (draft). National Aeronautics and Space Administration unpublished report, 1974.

Collins, Stewart A., et al. "First Voyager View of the Rings of Saturn." *Nature* 288 (1980), 439–442.

Cooper, Henry S. F., Jr. "A Reporter at Large: Imaging Saturn." *The New Yorker* August 24 (1981), 39–81.

Cuzzi, Jeffrey N., Jack Lissauer, and Frank H. Shu. "Density Waves in Saturn's Rings." *Nature* 292 (1981), 703–707.

Eshleman, V. R., G. L. Tyler, J. D. Anderson, G. Fjeldbo, G. S. Levy, G. E. Wood, and T. A. Croft. "Radio Science Investigations with Voyager." *Space Science Reviews* 21 (1977), 207–232.

Evans, David R., James W. Warwick, Jeffrey B. Pearce, Thomas D. Carr, and John J. Schauble. "Impulsive Radio Discharges near Saturn." *Nature* 292 (1981), 716–718.

Lissauer, Jack, Frank H. Shu, and Jeffrey N. Cuzzi. "Moonlets in Saturn's Rings?" *Nature* 292 (1981), 707–711.

Morrison, David. *Voyages to Saturn*. Washington, DC: National Aeronautics and Space Administration, 1982.

Smith, B. A., G. A. Briggs, G. E. Danielson, A. F. Cook, M. E. Davies, G. E. Hunt, H. Masursky, L. A. Soderblom, T. C. Owen, C. Sagan, and V. E. Suomi. "Voyager Imaging Experiment." *Space Science Reviews* 21 (1977), 103–127.

Smith, Bradford A., et al. "Encounter with Saturn: Voyager 1 Imaging Science Results." *Science* 212 (1981), 163–191.

Tyler, G. L., V. R. Eshleman, J. D. Anderson, G. S. Levy, G. F. Lindal, G. E. Wood, and T. A. Croft. "Radio Science Investigations of the Saturn System with Voyager 1: Preliminary Results." *Science* 212 (1981), 201–206.

Warwick, J. W., et al. "Planetary Radio Astronomy Observations from Voyager 1 near Saturn." *Science* 212 (1981), 239–243.

Chapter 8

Brahic, André, and Bruno Sicardy. "Apparent Thickness of Saturn's Rings." *Nature* 289 (1981), 447–450.

Cuzzi, Jeffrey. "Planetary Ring Systems." *Reviews of Geophysics and Space Physics* 21 (1983), 173–186.

Cuzzi, Jeffrey N., Joseph A. Burns, Richard R. Durisen, and Patrick M. Hamill. "The Vertical Structure and Thickness of Saturn's Rings." *Nature* 281 (1979), 202–204.

Esposito, L. W., J. N. Cuzzi, D. R. Evans, J. B. Holberg, E. A. Marouf, and C. C. Porco. "Saturn's Rings: Structure, Dynamics, and Particle Properties." In *Saturn*, edited by T. Gehrels and M. Matthews. Tucson: University of Arizona Press (1984).

Evans, D. R., J. H. Romig, and J. W. Warwick. "Saturn Electrostatic Discharges: Properties and Theoretical Considerations." *Icarus* 54 (1983), 267–279.

Gurnett, D. A., E. Grun, D. Gallagher, W. S. Kurth, and F. L. Scarf. "Micron-Sized Particles Detected near Saturn by the Voyager Plasma Wave Instrument." *Icarus* 53 (1983), 236–254.

Lane, Arthur L., Charles W. Hord, Robert A. West, Larry W. Esposito, David L. Coffeen, Makiko Sato, Karen E. Simmons, Richard B. Pomphrey, and Richard B. Morris. "Photopolarimetry from Voyager 2: Preliminary Results on Saturn, Titan, and the Rings." *Science* 215 (1982), 537–543.

Marouf, Essam A., G. Leonard Tyler, Howard A. Zebker, Richard A. Simpson, and Von R. Eshleman. "Particle Size Distributions in Saturn's Rings from Voyager 1 Radio Occultation." *Icarus* 54 (1983), 189–211.

Michaux, C. "Saturn's E-Ring and the Particle Hazard to Voyager Spacecraft." Final Report on Jet Propulsion Laboratory, contract BP-713111. Photocopy.

Morrison, David. *Voyages to Saturn*. Washington, DC: National Aeronautics and Space Administration, 1982.

Pang, Kevin D., Charles C. Voge, Jack W. Rhoads, and Joseph M. Ajello. "The E-Ring of Saturn and Its Satellite Enceladus." *Journal of Geophysical Research*. (In press.)

Pollack, James B., and Jeffrey N. Cuzzi. "Rings in the Solar System." *Scientific American* 245 (1981), 105–129.

Smith, Bradford, et al. "A New Look at the Saturn System: The Voyager 2 Images." *Science* 215 (1982), 504–537.

Stone, E. C., and E. D. Miner. "Voyager 2 Encounter with the Saturnian System." *Science* 215 (1982), 499–504.

Tyler, G. Leonard, Essam A. Marouf, Richard A. Simpson, Howard A. Zebker, and Von R. Eshleman. "The Microwave Opacity of Saturn's Rings at Wavelengths of 3.6 and 13 cm from Voyager 1 Radio Occultation." *Icarus* 54 (1983), 160–188.

Weidenschilling, Stuart J., Clark R. Chapman, Donald R. Davis, and Richard Greenberg. "Ring Particles: Their Collisional Interactions and Physical Nature." In *Planetary Rings*, edited by R. Greenberg and A. Brahic. Tucson: University of Arizona Press (1984).

Chapter 9

Brahic, A., ed. *Planetary Rings*. I.A.U. No. 75. Toulouse: CNES and CEPADUES Editions, 1984.

Brush, Stephen G., C. W. F. Everitt, and Elizabeth Garber, eds. *Maxwell on Saturn's Rings*. Cambridge, MA: MIT Press, 1983.

Goldreich, Peter, and Scott Tremaine. "The Dynamics of Planetary Rings." *Annual Reviews of Astronomy and Astrophysics* 20 (1982), 249–283.

Greenberg, R., and A. Brahic, eds. *Planetary Rings*. Tucson: University of Arizona Press (1984).

Hartmann, William K. *Moons and Planets*, 2nd ed. Belmont, CA: Wadsworth, 1983.

Chapter 10

Brecher, K., A. Brecher, P. Morrison, and I. Wasserman. "Is There a Ring around the Sun?" *Nature* 282 (1979), 50–52.

Dobrovolskis, Anthony R. "Where Are the Rings of Neptune?" *Icarus* 43 (1980), 222–226.

Elliot, J. L., D. J. Mink, J. H. Elias, R. L. Baron, E. Dunham, J. E. Pingree, R. G. French, W. Liller, P. D. Nicholson, T. J. Jones, and O. G. Franz. "No Evidence of Rings around Neptune." *Nature* 294 (1981), 526–529.

Guinan, E. F., C. C. Harris, and F. P. Maloney. "Evidence for a Ring System of Neptune." *Bulletin of the American Astronomical Society* 14 (1982), 658.

Wilford, John Noble. "Data Show Two Rings Circling Neptune." *New York Times* June 10 (1982), B10.

Chapter 11

French, R. G., J. L. Elliot, and D. A. Allen, "Inclinations of the Uranian Rings." *Nature* 298 (1982), 827–829.

Nock, Kerry. "Rendezvous with Saturn's Rings." Jet Propulsion Laboratory unpublished report.

Appendix

Beatty, J. Kelly, Brian O'Leary, and Andrew Chaikin, editors. *The New Solar System* (2nd ed.). Cambridge: Cambridge University Press, 1982.

Index